Regional Problems of Earth Remote Sensing (RPERS 2019)

E3S Web of Conferences Volume 149 (2020)

Krasnoyarsk City, Russia
10 - 13 September 2019

Editors:

V.B. Kashkin
V. Kharuk
E. Loupian
Y.A. Maglinets
M.V. Noskov
L.G. Sverdlik
G.M. Tsibulskii

ISBN: 978-1-7138-0688-2

Printed from e-media with permission by:

Curran Associates, Inc.
57 Morehouse Lane
Red Hook, NY 12571

Printed with permission by Curran Associates, Inc. (2020)

For additional information, please contact EDP Sciences – Web of Conferences at the address below.

EDP Sciences – Web of Conferences
17, Avenue du Hoggar
Parc d'Activité de Courtabœuf
BP 112
F-91944 Les Ulis Cedex A
France

Phone: +33 (0) 1 69 18 75 75
Fax: +33 (0) 1 69 28 84 91

contact-edps@webofconferences.org

Additional copies of this publication are available from:

Curran Associates, Inc.
57 Morehouse Lane
Red Hook, NY 12571 USA
Phone: 845-758-0400
Fax: 845-758-2633
Email: curran@proceedings.com
Web: www.proceedings.com

TABLE OF CONTENTS

Histogram comparison crosscalibration method applied for evaluating the quality of "Canopus-v-ir" data

Mikhail Burcev[1], *Alexandr* Kashnitskiy[1], *Evgeny* Loupian[1], *Alexey* Mazurov[1], *Ilya* Pustynskiy[2], and *Egor* Kholodov[2]

[1]Space Research Institute, Russian Academy of Sciences, Moscow, Russia
[2]Far-Eastern center SRC "Planeta", Khabarovsk, Russia

Abstract. Regular onboard calibration of satellite data is of great practical importance for the processing of remote sensing information. For flight cross-calibration it is proposed to use a geostationary satellite as a reference to calibrate polar orbiting satellites data on the example of the "Canopus-V-IR" and Himawary satellites. Authors use data from center for collective use "IKI-Monitoring", united data catalogue of Planeta Research Centers for Space Hydrometeorology. The work has been done on theme «Monitoring»

Modern remote sensing devices are the measuring instruments that provide data used in in various tasks of global changes monitoring in the earth's surface and determining dangerous phenomena. The accuracy and stability of the calibration characteristics of satellite instruments are important for the development and sustainable automatic application of processing algorithms. During the operation of spacecraft, it is necessary to conduct regular flight calibration of instruments, for which various techniques are used [1-4].

The method proposed for use is based on the comparison of histograms of images from the same Earth's surface scene by instruments from different spacecraft having similar spectral characteristics and made with the same view geometry and at the same time [5]. One of the devices will be considered a reference, with good calibration. We will try to calibrate other devices according to that one.

It seems natural to assume that the sensitivity of the device has no inversion and that the magnitude of the signal recorded by the device is proportional to the value of the radiation flux received from the Earth, i.e. a brighter object on earth corresponds to a stronger signal recorded by the device. In this case, all objects with the same brightness should have the same area in the images from different devices, which in terms of the image means the same histogram value in the area corresponding to the brightness of the objects.

Thus, constructing the correspondence of the partial areas of the histograms, we obtain an unambiguous correspondence of the brightnesses of one image to the brightnesses of another, or in other words, the correspondence between the recorded values of one device and the registered values of another. With a good calibration of the reference device, we get a calibration table for the recorded values of another device.

One of the difficulties of such a scheme is the condition for simultaneous shooting with the same geometry. Nevertheless, when analyzing images of orbital and geostationary spacecraft, one can find approximate fulfillment of both the conditions of the geometry of the view and the conditions of simultaneity with an accuracy of less than 5 minutes, and more accurately with planning the session.

The study was conducted using data access and processing services “VEGA-Science” [6]. For the Canopus-V-IR data, scenes near to Himawari subsatillite point were provided by “VNIIEM Corporation” JC for April 12-15, 2019. Himawari data processed at the Far Eastern center SRC "Planeta" [7] were used as a reference.

The geometry of the both views of the scene from the geostationary and polar-orbiting satellite is shown in Fig. 1. On the images taken from two satellite the areas are selected to satisfy the geometry with difference of up to 5 degrees of satellites zenith angles.

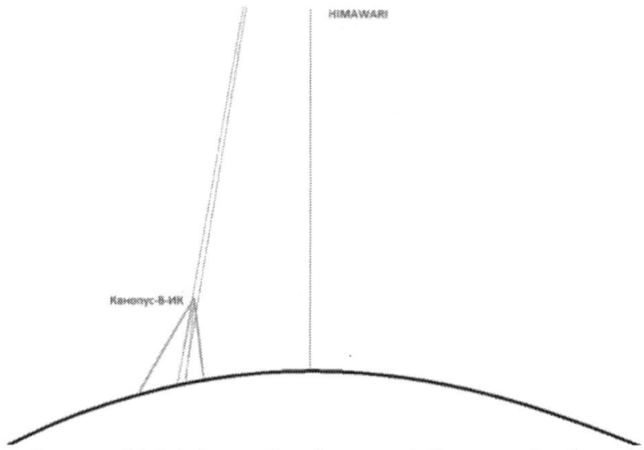

Fig 1. . Geometry diagram of joint observation from geostationary and polar-orbiting satellite.

Images by Canopus-V-IR are averaged to the images resolution of the AHI HIMAWARI satellite (these data are used as reference data). Further, according to the histograms of images using the method described above the calibration tables are calculated for the data "Canopus-_V-IK", MSU-IK-SRM device.

The spectral characteristics of the devices in the bands of 8.6 μm and 3.7 μm are shown in Fig. 2

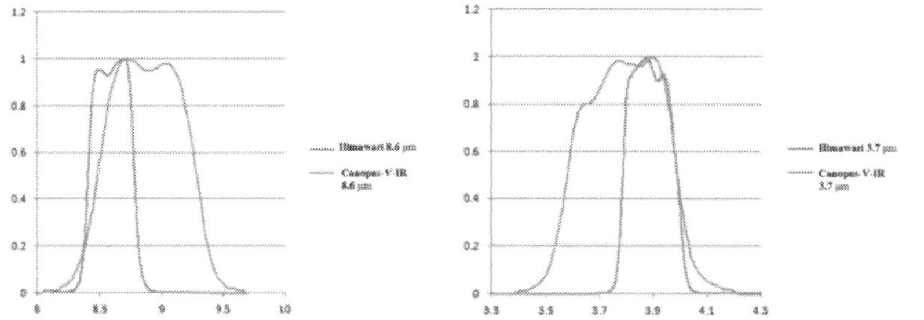

Fig 2 Spectral response sensitivities.

An example of a pair of images used for cross-calibration with a selected fragment for histogram analysis is shown in Fig. 3. The corresponding histograms and data conversion graph (calibration table) are shown in Fig. 4.

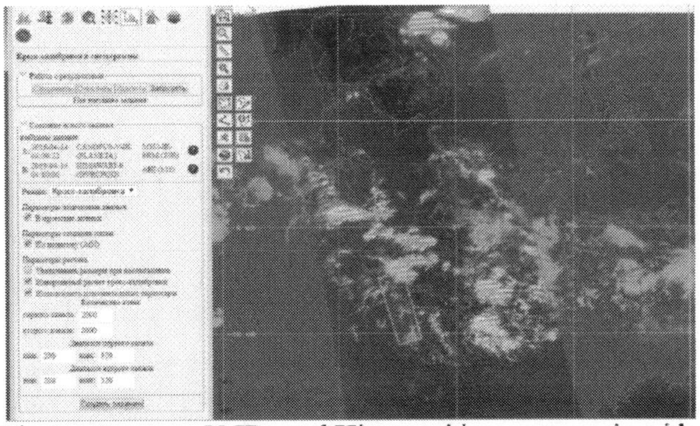

Fig 3. «Canopus-V-IR» and Himawari images mosaic with region for histogram analysis

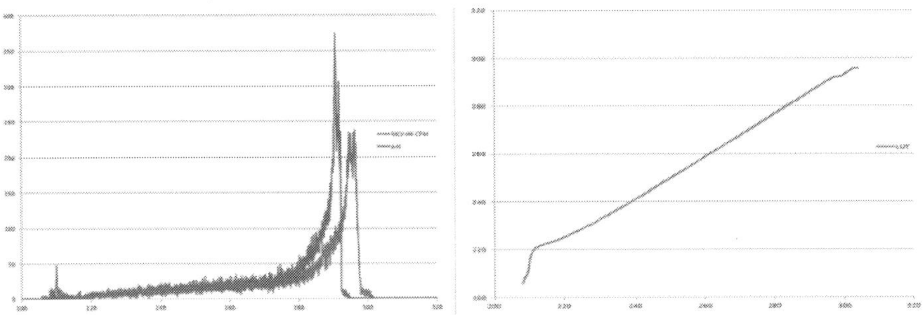

A. B.

Fig 4 Histogram.(A). Calibration table (B)

A number of factors affecting the proposed procedure should be noted. The Canopus-V-IR data has obvious significant noise in the image (Fig. 5). This fact leads to errors in the construction of calibration tables. In particular, the apparent periodic strips in the warm zone (dark and irregular shades of the strip in the images Fig.56 a, b) overestimate part of the histogram of this region, this affects the conversion table in the upper right part of the graph of the calibration curve with higher than real temperature values and greater error in this area. Irregular values of other failures in the image (over the entire range of values) give corresponding errors to the conversion table, especially in places of the histogram with small values (in a colder area).

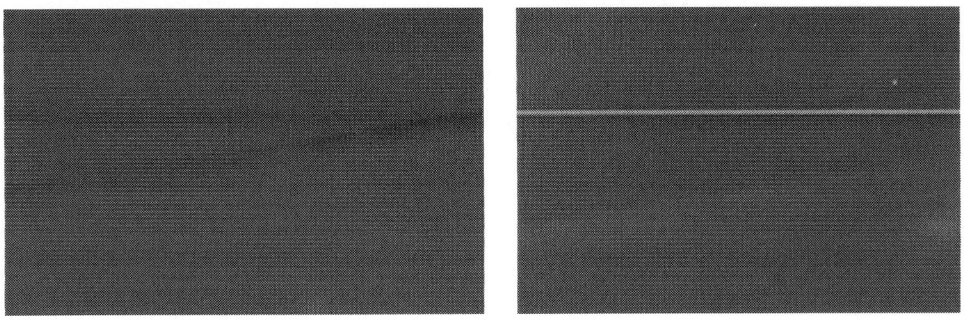

A B

Fig 5. Noise strips over images.

Calibration tables are constructed and averaged over a set of images (the graphs in abstract counts are shown in Fig. 6a and the result of the averaged table in terms of degrees K in Fig. 6b.

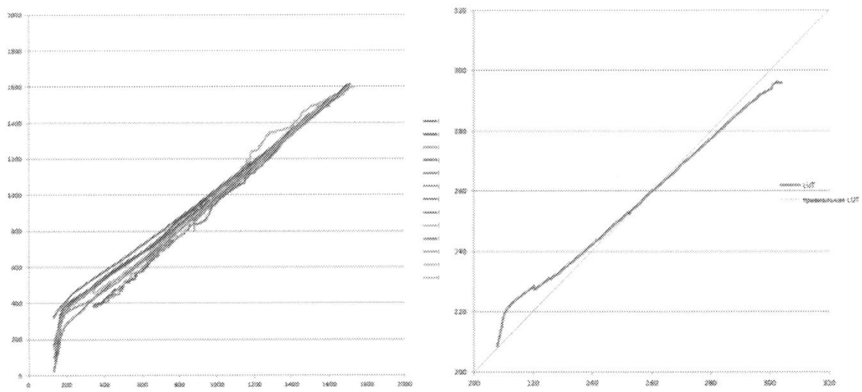

A B

Fig 6.Calibration tables: set (A) and average (B).

The error (standard deviation) is from 1 (warm region) to 2.5 degrees K (cold region).

An almost linear course of the calibration curve should be noted, and only in the temperature range of 220 K and lower does the sensitivity drop, which appears in the histogram as a small upward surge in its values.

Sufficiently large deviations from the average are not associated with a possible sharp change in weather conditions in the considered regions within 5 minutes. For all data sets, the same conversion tables can be constructed for successive observations from the Himawari satellite. An example of the table graph for this case is shown in Fig.7. It can be seen that it differs from the trivial linear transformation of the image itself, but not as much as the table graphs for the Canopus-V-IR-Himawari image pairs.

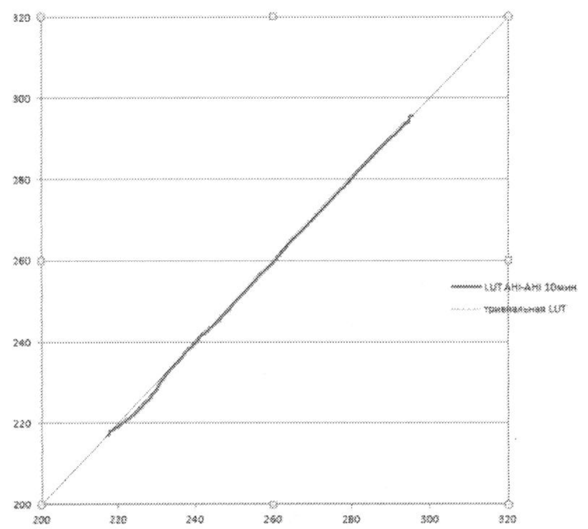

Fig 7 10 min delayed AHI images self-calibration table.

Thus, the error in the construction of refined calibration characteristics can be explained by the noise of the data. At the same time, the relatively close course of the calibration curves constructed using the above technique allows one to conclude that the Canopus-V-

IR images after eliminating noisy data could be calibrated with using this cross-calibration technique.

The existing image set also contains data in a 3.7 μm spectral band. But they are so noisy (Fig.8) that, due to the above considerations, this technique is not applicable for performing cross-calibration.

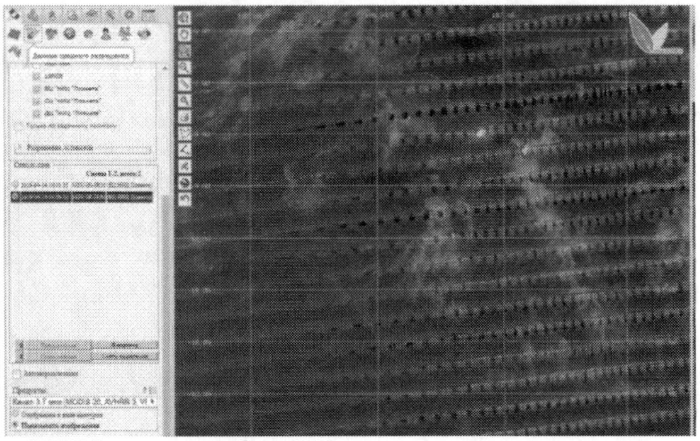

Fig 8 Canopus-V-IR image example at 3.7 μm.

Authors use data from center for collective use "IKI-Monitoring", united data catalogue of Planeta Research Centers for Space Hydrometeorology. The work has been done on theme «Monitoring»

1. G. Zhou, C. Li, T. Yue, L. Jiang, N. Liu, Y. Sun, M. Li An overview of in-orbit radiometric calibration oftypical satellite sensors, Remote Sensing and Spatial Information Sciences, 2015, Vol. XL-7/W4, 2015 International Workshop on Image and Data Fusion (2015)
2. G. Chander, T. J. Hewison, N. Fox, X. Wu, X. Xiong, J. William Overview of Intercalibration of Satellite Instruments, IEEE Transactions on geoscience and remote sensing (2013)
3. LiuJ.-J., Z. Li, Y.-L. Qiao, Y.-J Liu, Y.-X. Zhang A new method for cross-calibration of two satellite sensors, IEEE Int. J. Remote sensing (2004)
4. S. Sterckx, S. Adriaensen, W. Dierckx, M. Bouvet In-Orbit Radiometric Calibration and Stability Monitoring of the PROBA-V Instrument, Remote Sens. (2016)
5. A.V. Kashnitskiy, E.A. Loupian, A.A. Mazurov, A.M. Matveev The Possibility of Automatic Intercalibration for Russian Satellites by Histogram Analysis // "Information Technologies in Remote Sensing of the Earth-RORSE 2018". IKI RAS (2019)
6. E.A. Loupian, A.A. Proshin, M.A. Bourtsev, I.V. Balashov, S.A. Bartalev, V. Yu. Efremov, A.V. Kashnitskiy, A.A. Mazurov, A.M. Matveev, O.A. Sudneva, I.G. Sychugov, V.A. Tolpin, I.A. Uvarov IKI center for collective use of satellite data archiving,processing and analysis systems aimed at solving the problems of environmental study and monitoring // Actual Problems of Remote Sensing of the Earth from Space (2015)
7. E.A. Loupian, O.E. Milexin, V.N. Antonov, L.S. Kramareva, M.A. Bourtsev, I.V. Balashov, V.A. Tolpin, V.I. Solovyev System of operation of joint information resources based on satellite data in the Planeta Research Centers for Space Hydrometeorology // Russian Meteorology and Hydrology (2014)

Development of the unit for testing the methods of remote computerized monitoring of environmental changes in the "IKI-Monitoring" center for collective use

Evgeny Loupian[1,*], *Viktor* Savorskiy[1,2], *Aleksandr* Kashnitskii[1], *Dmitrii* Kobets[1], *Konstantin* Sen'ko[1], *Ivan* Balashov[1] и *Ivan* Uvarov[1]

[1]Space Research Institute of the Russian Academy of Sciences, 117997 Moscow, Russia
[2]Kotelnikov Institute of Radioengineering and Electronics of the Russian Academy of Sciences, Fryazino Department 141190 Fryazino, Moscow region, Russia

Abstract. The impact of hazardous waste on people and the environment from agricultural, industrial, military and mining activities form one of the most significant global environmental challenges. These wastes often include heavy metals, acid precipitates (sulfur oxide solutions), hydrocarbons, and other organic chemicals. Creation of the tools of the VEGA-Science service (http://sci-vega.ru/) to control the state of vegetation for the monitoring the impact of technogenic waste and dumps sources (TWDS) on the environment is the main objective of this work. To achieve this goal, proposals for the modernization and development of the VEGA-Science services have been defined to provide the organization and analysis of long-term variability of the vegetation cover caused by the influence of TWDS.

1 Introduction

The paper [1] gives an analytical review describing the mechanisms of influence of typical harmful TWDS on the vegetation covers and proposes methods and approaches that allow to identify changes in the vegetation cover caused by the influence of TWDS. This work results in the statement of proposals for making up the comprehensive list of parameters for monitoring forest vegetation cover based on satellite observations, including the use of spectral indices for these purposes. The obtained results allowed to substantiate the use of the funds and resources of IKI-Monitoring [2] as a platform for the development of tools to control the impact of TWDS on the environment. The points of the paper [1] underwent the further development resulting in the hereby article.

The main objective of this work is the creation of the tools in framework of the VEGA-Science system for assessment of the state of the vegetation cover for monitoring the impact of technogenic waste and dumps sources (TWDS) on the environment. To achieve this goal, proposals for the modernization and development of the VEGA-Science services

* Corresponding author: evgeny@d902.iki.rssi.ru

have been defined to provide the organization and analysis of long-term variability of the vegetation cover caused by the influence of TWDS.

The main methodological approach in the work is the object-oriented design method, based on the clarification of user requirements and the development of software services that meet these target requirements.

2 Requirements for source satellite and auxiliary data for the assessment of the impact of TWDS on the vegetation cover

The peculiarity of the algorithms for detecting the impact of TWDS is the need to consistently, in a wide range of hydrometeorological parameters, detect the disturbance of vegetation against the background of strong interannual, seasonal and intra-seasonal variability. This necessity is determined by the peculiarities of temperature and humidity conditions, which develop differently in each of the annual cycles, which leads to significant differences in the phenological cycles of plant development. Therefore, the detection of disturbance caused by TWDS, involves the use of such methods of monitoring the state of vegetation, which are resistant to changes in temperature and humidity conditions, in order to ensure long-term monitoring of control areas of forest cover for 10-20 years for at least one of the types of disturbance described in the paper [1].

The need to detect the influence of TWDS at significantly different temperature and humidity observation conditions form the following number of requirements for the list of the source and auxiliary data that ensure stable operation of the algorithms for detecting forest canopy disturbance caused by TWDS:

1) TWDS description database (coordinates, type, formation period, polluting factor),
2) database of the forest plots (allotments) having close characteristics and taxonomy (main breeds, stock, bonitet) to the ones in the areas of TWDS,
3) database for search and selection of forest areas (allotments) having similar weather and climatic characteristics (temperature, humidity, soil moisture, wind speed and direction) to the ones in the areas of TWDS,
4) medium and high spatial resolution historical satellite observations data archive for selected control areas of the forest,
5) medium and high spatial resolution real time satellite observations data archive for selected forest areas.

3 Requirements for the services of detection of the TWDS impact on the vegetation cover based on the satellite observations

As follows from the analysis, in order to automatically detect the TWDS impact on forest vegetation cover, an automated system for remote monitoring of vegetation cover (SRMV) in the TWDS areas should be launched, being able to provide the following capabilities:

1) service for definition of the list of the forest sites (allocations) having similat taxonomy (basic breeds, stock, bonitet) to the ones in the areas of TWDS,
2) service for definition of the list of forest areas (allotments) having similar weather and climatic characteristics (temperature, humidity, soil moisture, wind speed and direction) to the ones in the areas of TWDS,
3) service for making up the pairs ("observed site" - "reference site") based on the similarity in taxonomy and weather and climatic conditions for each observed TWDS (taking into account the wind rose and proximity of the sites to TWDS),

4) service for calculation of the spectral indices characterizing the state of the vegetation cover, including NDVI, AVFI1600, DWSI5, EVI, NDNI, OSAVI, PEI, TCARI (see the paper [1] and references) and TIEHD [3, 4], based on medium and high spatial resolution to historical and current satellite observations for pairs ("observed site" - "reference site") of forest plots (allotments);
5) service of estimation of difference in spectral indices for the pairs of the sites ("observed site" - "reference site") having the same weather and climatic conditions and having the same taxonomy at the beginning of monitoring,
6) service for analysis of phenological seasonal cycles of pairs of spectral indices for pairs of sites ("observed site" - "reference site") having the same taxonomy at the beginning of monitoring,
7) service for detection and monitoring of long-term cumulative trends of spectral index differences for pairs ("observed site" - "reference site"),
8) service for distinguishability assessment (statistical measure of distinguishability) for pairs ("observed site" - "reference site") having the same weather and climatic conditions, in particular, using the Jeffries-Matsushita distance [5].

All the above algorithms and services should be provided with remote access user web interfaces [6].

4 The "IKI-Monitoring" Center for Collective Use, the information platform to provide the potential capabilities and resources for solving the problems of monitoring of dumps of technogenic waste

The requirements listed in sections 2, 3 are substantially satisfied by the information tools developed and implemented in the information system of the IKI-Monitoring CCU (http://ckp.geosmis.ru/). Thus, the VEGA-Science services and tools can be used as prototypes during the development of tools for remote access to the remote forest monitoring data in order to detect forest disturbance due to the impact of TWDS. They provide:

1) accurate geographical reference of the source satellite monitoring data and derived data products, including spectral indices,
2) creation of the TWDS database,
3) automatic and semi-automatic definition of the polygon network,
4) calculation of statistical parameters of the source satellite monitoring data and derived data products, including spectral indices for selected polygons,
5) creation of the database of taxonomically homogeneous forest areas,
6) recognition of homogeneous weather and climatic zones according to long-term hydrometeorological observations,
7) making of spectral index maps, including the ones edited by users in the online mode,
8) creation of the database of seasonal phenological profiles of source satellite monitoring data and derived data products for selected polygons,
9) creation of decadal, monthly and seasonal average values of spectral indices based on long-term satellite monitoring of the polygon network.

The availability of such prototype tools ensures the rapid successful implementation of the TWDS SRMV as a unit of the VEGA-Science system. This is also facilitated by the operation of the IKI-Monitoring center, which includes VEGA-Science systems, information resources of long-term satellite monitoring and related information. The prototype of the specialized unit for remote monitoring of vegetation cover in TWDS areas is currently in operation, providing the implementation of TWDS SRMV as a part of the system. An example of the interfaces that provides the operation of this unit is shown in Fig. 1.

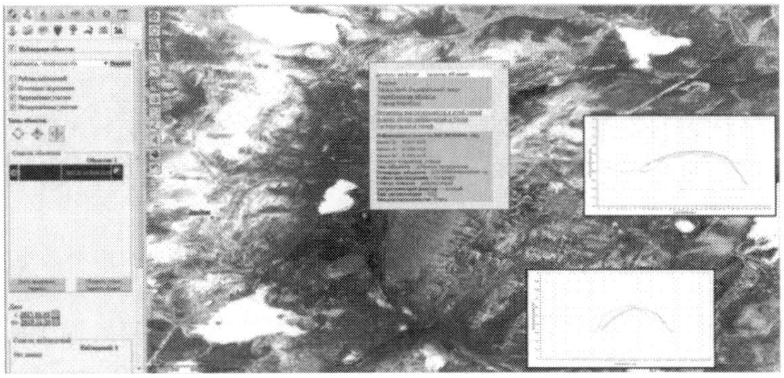

Fig. 1 Interfaces of the specialized unit for remote monitoring of vegetation cover in the TWDS areas. Data analysis in the “Kabashmed” plant dumps area (Chelyabinsk region).

Conclusions

Proposals for the modernization and improvement of VEGA-Science services, which are necessary for the organization and analysis of long-term variability of vegetation caused by the influence of TWDS, are formulated. In order to test the feasibility of the proposal implementation, a working prototype of a specialized unit for remote monitoring of vegetation cover in the TWDS regions was created as part of the VEGA-Science system.

The study was supported by RFBR in framework of the scientific project No. 18-29-24121 MK.

References

1. V.P. Savorskiy, E.A. Loupian, V.I. Gornyy, D.M. Ermakov, O.Yu. Panova, A.V. Konstantinova, Sovremennye problemy distantsionnogo zondirovaniya Zemli iz kosmosa, **16**(6), (2019). (in print)
2. E.A. Loupian, A.A. Proshin, M.A. Bourtsev, A.V. Kashnitskii, I.V. Balashov, S.A. Bartalev, A.M. Konstantinova, D.A. Kobets, A.A. Mazurov, V.V. Marchenkov, A.M. Matveev, M.V. Radchenko, I.G. Sychugov, V.A. Tolpin, I.A. Uvarov, Sovremennye problemy distantsionnogo zondirovaniya Zemli iz kosmosa, **16**(3), 151-170 (2019). DOI: 10.21046/2070-7401-2019-16-3-151-170.
3. V.I. Gornyy, S.G. Kritsuk, I.Sh. Latypov, V.N. Khramtsov, Sovremennye problemy distantsionnogo zondirovaniya Zemli iz kosmosa, **10**(4), 201-212 (2013).
4. V.I. Gornyy, A.V. Kiselev, S.G. Kritsuk, I.Sh. Latypov, A.A. Tronin, Sovremennye problemy distantsionnogo zondirovaniya Zemli iz kosmosa, **16**(4), 124-136 (2019). DOI: 10.21046/2070-7401-2019-16-4-124-136.
5. J.A. Richards, X. Jia, Remote Sensing Digital Image Analysis. Berlin Heidelberg: Springer, (2006), 439 p.
6. E.A. Loupian, 1. V.P. Savorskiy, Yu.I. Shokin, A.I. Aleksanin, R.R. Nazirov, I.V. Nedolugko, O.Yu. Panova, Sovremennye problemy distantsionnogo zondirovaniya Zemli iz kosmosa, **9**(5), 21-44 (2012).

Meteorological data analysis capabilities in the "VEGA constellation" information systems family

Ivan Uvarov, *Viktor* Marchenkov, *Vladimir* Tolpin

Space Research Institute, 117997, 84/32, Profsoyuznaya str., Moscow, Russia

Abstract. Satellite monitoring information systems require joint analysis of heterogeneous data, including meteorological data. To solve these problems, IKI RAS developed a technology for automatic acquisition, storage, presentation and analysis of meteorological data. The archive of meteorological data in IKI RAS is maintained within the Center for Collective Use "IKI-Monitoring". The data is being continuously updated from a number of sources, primarily with the data obtained by the NCEP model. The data is available for visualization on the map in various ways. Plotting provides time series analysis in comparison to other variables such as vegetation indices. The developed technologies have been applied in development of a number of specialized information systems of satellite monitoring of natural and anthropogenic objects.

The development of specialized thematic information systems for satellite monitoring involves the possibility of joint analysis of remote sensing data with other types of data, one of which is meteorological data. [1,2] the range of applications of meteorological data is quite wide: assessment of the risks of natural fires, the impact of drought on plant development, monitoring the spread of ash plumes during volcanic eruptions, the study of processes on the ocean surface, and much more.

To solve these problems, RAS IKI has developed the technology for automatic acquisition, storage, visualization and analysis of weather data. [3]

The data used in the information systems include forecast and historical data obtained from various sources, which differ significantly in the set of available variables, time resolution and spatial localization.

The source data is usually localized either at the points of weather stations (including the measured data), or in the nodes of the regular grid. Data that has been statistically processed can also be presented in terms of units of spatial division, for example, regions or districts.

The data archives obtained using the Climate Forecast System Version 2 (CFSv2) NCEP model are constantly updated (https://www.ncep.noaa.gov/). the Spatial resolution of the data is 0.25° and the temporal resolution is 3 hours. The archives of the RAS IKI contain more than two dozen indicators of the state of the atmosphere at the Earth's surface, as well as vertical temperature profiles on various Isobaric surfaces.

Various methods can be used to display weather data in the map interface. Depending on the nature of the data, isolines, cartograms, and icons are used. Isolines can be used for data of any spatial localization, but are best suited for data based on a regular grid. When analyzing data localized at weather stations, it is reasonable to use icons. In this case, the location of the weather station shows rectangular icons, characterized by the color and numerical value of the displayed value. Simultaneous analysis of several values is possible by applying complex icons at the location of the weather station, where the numerical values, provided with color, are arranged in a column. Data can be visualized in the context of territorial divisions. In this case, the mapping method is used. Indicators can be presented in both absolute and relative terms.

Analysis of the temporal dynamics of meteorological variables is possible for objects of different types: individual points, objects of observation, weather stations, areas. A special user interface presents data in the form of graphs. This makes it possible to jointly analyze the dynamics of various meteorological indicators and the dynamics of other variables, such as vegetation indices obtained from a time series of satellite observations. A comparative analysis of the intra-seasonal dynamics of different years is available, as well as a comparison of the climatic characteristics of different objects.

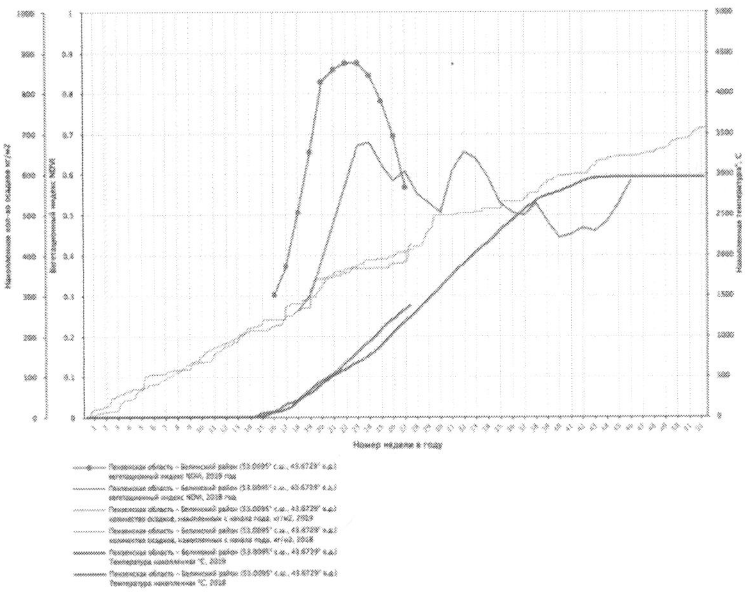

Fig. 1. Time series analysis in the VEGA-Science system. Time series of accumulated temperature and precipitation in comparison with the NDVI index.

Derived indicators are the result of statistical processing of climate data and are designed to detect weather anomalies on certain natural or man-made objects. In particular, any of the key indicators, such as temperature or precipitation, apply transformations, such as the calculation of the multi-year average for specific day of the year, calculation of the multi-year monthly average, calculation of current deviations from the multiyear average, calculation of accumulated value since the beginning of the year, determination of regional norm, windowing etc. It is possible to calculate temperature indices, the hydrothermal coefficient. The combination of such transformations is intended to help identify the most significant factors in the analysis of the temporal dynamics of climate data.

Conclusion. The proposed technologies and approaches allow us to develop tools for analyzing meteorological data. This has been used in a number of information systems for monitoring natural and man-made objects. The operation of subsystems for archiving and access to weather data is provided by the facilities of the "IKI-Monitoring" Center for Collective Use. [4]

References

1. Tolpin V.A., Balashov I.V., Loupian E.A., Savin I.Yu. "VEGA" satellite service // Earth from Space. 2011. Issue 9, spring,. P. 32-37.
2. Loupian E.A., Bartalev S.A., Tolpin V.A., Krasheninnikova Yu.S., Oksyukevich A.Yu. Opportunities and experience of using the VEGA satellite service in the organization of regional remote monitoring // Regional problems of remote sensing of the Earth: proceedings of the international scientific conference / ed. E.A. Vaganov. Krasnoyarsk: SFU. ISBN 978-5-7638-3109-2, 2014. C. 3-6.
3. Uvarov I.A., Khalikova O.A., Balashov I.V., Bourtsev M.A., Loupian E.A., Matveev A.M., Platonov A.E., Proshin A.A., Tolpin V.A., Krasheninnikova Yu. S. Meteorological data management in framework of the satellite monitoring information systems // Actual Problems of Remote Sensing of the Earth from Space. 2013. Vol.10. № 2. P. 30-45.
4. Loupian E.A., Proshin A.A., Burtsev M.A., Balashov I.V., Bartalev S.A., Efremov V.Yu., Kashnitskiy A.V., Mazurov A.A., Matveev A.M., Sudneva O.A., Sychugov I.G., Tolpin V.A., Uvarov I.A. IKI center for collective use of satellite data archiving, processing and analysis systems aimed at solving the problems of environmental study and monitoring, // Actual Problems of Remote Sensing of the Earth from Space, 2015, 12(5), pp. 263–284.

Interface Model for Spatial Objects Monitoring Task Setting

Ruslan V. Brezhnev[1,*], *Yurii A.* Maglinets[1], *Alena V.* Kornilaeva[1] and *Evgenii A.* Maltsev[2]

[1] Siberian Federal University, The Institute of Space and Information Technology, Kirenskogo st. 26, Krasnoyarsk, 660074 Russia
[2] Skolkovo Institute of Science and Technology, Bolshoy Boulevard 30, bld. 1, Moscow, 121205, Russia

Abstract. The research is devoted to the study of human-computer interaction in the context of interaction of end users (EU) and multi-purpose remote monitoring systems (MRMS) for spatial objects. The study is aimed at the formation of conceptual aspects of building graphical interface for setting monitoring tasks. It defines a global structure for communication between end users and multi-purpose remote monitoring systems (MRMS) in the form of a dialogue and target parameters for the quality of the dialogue tools. The task setting interface has been tested in practice on the basis of the ISIT SFU remote monitoring system. An example of task setting for grain crops growth monitoring has been reviewed.

1 Introduction

The research of "human - information system" communication is an integral part of information science. The research focuses on building communication between users and data processing systems (DP-systems) which require data processing chain management. DP-systems can include various processes depending upon the application of such systems which makes the development of a unified approach to dialogue interaction quite a challenging task.

The research focuses on the analysis of spatial objects remote monitoring systems used for Earth remote sensing data processing. At present, there is a growing demand for information products based on Earth remote sensing data [1]. The users of such systems are state administrative authorities and commercial entities, as well as individuals. Users from all these groups shall be considered end users, i.e. task setters and results consumers. The number of information requests from end users tends to increase which require certain interface solutions [2]. Processes for solving satellite monitoring tasks are often of combined type with one part of data processing operations automated and the other part requiring the involvement of programmers which substantially restricts task setting for end users. In connection herewith, the structuring of available data processing algorithms which

* Corresponding author: brejnev.ruslan@gmail.com

allow using unified dialogue patterns by the end user on its own shall be a well-grounded solution for remote monitoring systems development.

2 Global dialogue structure model

Dialogue shall be a process with an end user pursuing a certain goal which can be analyzed at various levels. The global structure [3] is the highest level for dialogue structure analysis. It shall be the same for the whole class of tasks solved and shall not depend on specific features of separate tasks solved. The global structure is characterized by a general sequence of interaction stages which can further specify dialogue model with the help of subject-specific and local structures' terms which very much depend on the type of the task solved.

Figure 1 contains global structure for spatial object monitoring task class.

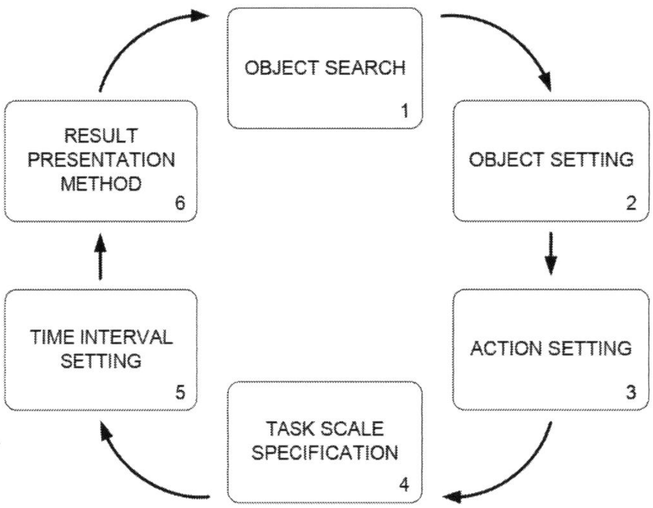

Fig. 1. Global dialogue structure model for spatial objects monitoring task setting.

The structure includes six stages which can be described as follows.

1. Object search shall be used to identify relevant information in the system data describing the target object. Data structure contains factual information about geospatial objects in the form of geospatial tiers, as well as data about the objects represented in the form of taxonomies [4]. The display of the searched spatial object contours on the map or a list of objects matching the request shall be a successful search result.

2. Object setting can be of two types. The first type shall include the use of the object found at the previous stage. In this case the user shall select object specialization and localize it with the help of geographical coordinates by highlighting the object at the map (Fig. 2). The second option is used in case the target object is absent in the system data structure. In this case the user can set the target object coordinates on its own by means of visual outlining.

3. Task setting (*or operations with the object*) shall depend on the specific type of the object defined at the previous stage. E.g. if the found or newly created object is a crop rotation field, the system shall suggest preset monitoring algorithms for vegetation heterogeneity or conditions based on vegetation indices. Actions shall be single or cyclic. Single actions shall include one processing iteration, e.g. object area calculation, while cyclic actions shall include a number of measurements at a specified time interval which

corresponds to the term "monitoring".

4. Task scale specification shall specify the information request and generally shall be an optional stage of the global dialogue structure. Task specification shall define the object with account of semantic characteristics. E.g. vegetation heterogeneity monitoring shall be required for the specified crop rotation field in case of wheat growing. Semantic characteristics shall be set in the form of classifiers based on taxonomy, with the spatial object connected by numerous relations to other objects of various types expanding the knowledge about it.

5. Time interval setting shall allow setting the time period for making operations with the target object. The following two aspects depending on step 1 and 2 shall be accounted for. Shall the object be found in the system, the database can contain information on its lifecycle, including on its active cycle. Shall there be no information about the object lifecycle in the database or shall the object be newly created in the process of task setting, then the user can set the monitoring interval at the timescale or define a set of time points on its own. The difference between the two approaches to time interval setting shall be the number of monitoring iterations which shall be directly proportionate to the number of the relevant Earth sensing data receipt iterations in the first case or shall depend on the number of time points and data availability as at the set date in the second case.

6. Setting the results presentation method shall define the method for measurement results storage and visual presentation to the user. In general, the results of measurements shall be represented in the form of diagrams or bar charts, and parameters analyses shall be presented in the form of schematic maps with map legend. Table sheets which can be downloaded as files can also be used for visual presentation of results.

3 Model testing in practice

A brief description of implementation of certain key steps of the global dialogue structure is given below (Fig. 2).

Fig. 2. Monitoring task setting interface for the specified spatial object.

Let us assume that the aim of the user is monitoring of changes in vegetation structure [5] of object "Crop rotation field No.158". Given that the information on the target object is present in the coordinates structure and the user has localized it in the coordinate space, the system shall suggest setting object specialization and selecting admissible action with the object, specifying the object parameters in the semantic space, e.g. by setting agricultural crop, as well as specifying the monitoring period from May 01, 2019 to September 30, 2019.

4 Conclusion

The analyzed dialogue structure model is based on ISIT SFU multi-purpose remote monitoring system (MRMS) and is currently at the testing and debugging stage. The key target parameters which define the quality of dialogue interaction and are prioritized in this research shall include the following: advanced system functionality with minimum amount of dialogue steps taken by the end user for task setting; flexible interaction which shall mean non-strict step sequence and can be present in various task setting dialogue paths; a possibility for the end user to specify the request more precisely in case of disagreement with the results which shall be considered by the system as the next problem solution iteration followed by the assessment of results.

The reported study was funded by Russian Foundation for Basic Research (No. 18-47-242002), Government of the Krasnoyarsk Territory, Krasnoyarsk Regional Fund of Science, to the research project: «The development of technology for creating intelligent information systems of object-oriented monitoring of territories based on remote sensing data».

References

1. E.A. Lupyan, I.V. Balashov, M.A. Burtsev, V.Yu. Efremov, A.V. Kashnitsky etc., *Conference: Modern problems of Earth remote sensing from Space*, **12**, 53 (2015)
2. V.I. Drakin, E.V. Popov, A.B. Preobrazhenskiy, *End users interaction with data processing systems* (M. Radio and Communication, 1988)
3. C. Traynor, M.G. Williams, *Interactive Technologies*, 115 (2001)
4. Yu. A. Maglinets, K. V. Raevich, G. M. Tsibulskii, *Procedia Engineering*, **201**, 331 (2017)
5. R.V. Brezhnev, Yu.A. Maglinets, K.V. Raevich, G.M. Tsibulski, *CEUR Workshop Proceedings*, **2210**, 316 (2018)

Application of remote methods for landscape taxation of forest-park forests

Oleg Artemjev[1] *and Larisa Baykalova*[2]
[1]Reshetnev Siberian State University of Science and Technology, 660037 Krasnoyarsk Russia
[2]Krasnoyarsk state agrarian University, 660049 Krasnoyarsk Russia

Abstract. Here provided are developed methodologies of landscape taxation of forest-park forests, based on the use of remote sensing methods. The methods of decoding by aerospace photographs of a walking road-path network and recreational forest digression are described. Methods, which are using terrestrial images, will improve the accuracy of assessment of landscape and architectural indicators: the closure of the canopy of the tree stand, the assessment of the aesthetics of the site, the passability of the terrain of the site, sanitary and hygienic assessment.

1 Introduction

Forest park forests are among the most valuable categories of forest and as a result, development to improve their inventory is relevant. The use of remote methods for taxation of forest-park forests will increase the accuracy of their assessment.

2 Suggested landscape inventory techniques

Forest park forest attendance is more dependent on the existence of a walking road-trail network.

The design of a walking road-path network should be given special attention, since errors in its design will not only reduce the attendance of the recreation object, but also cause unreasonable costs for the construction of those walking roads and paths that will be not enough visited.

To improve the design quality of the walking road-path network, the following methods is proposed: At the first stage, using the color spectrozonal aerial photographs or color spectral satellite photographs, the existing road-path network is decrypted. It is necessary to use large-scale aerospace photographs. Decryption of roads is straightforward and is carried out according to existing methods. Deciphering of walking trails is based on direct and

[1] Corresponding author: ol.artemjev2010@yandex.ru

indirect decoding attributes developed by the author using stereopairs of aerospace photographs [1]. Descriptions of the decryption features are shown in Table 1.

Table 1. Deciphering signs of walking trails.

Deciphering signs of walking trails				
Direct		Indirect		
Shape	Color	Spatial placement with stereoscopic perception	Proximity to public places of rest, public transport, houses, roads	Closed forest canopy
Thin, interconnected lines, in stands with high closeness - dashed lines	On dark gray to black lands covered with forest; on uncovered - the same as the surrounding area, but lighter	Trail trails are below the canopy of the stand	Most trails are near objects of mass visit	Most paths go through less densely forested stands.

In addition to decryption features, the use of "standards" is recommended, such as stereo pairs of images with an already decrypted path network.

The number of identified paths in aerial photographs of various scales is shown in the Table 2.

Table 2. The number of identified paths in aerial photographs of various scale.

Length paths (according to ground survey), km	Of these, identified on scale aerial photographs			
	1:5000		1:3000	
	КМ	% of the total length of paths	КМ	% of the total length of paths
85,06	79,45	93,4	80,89	95,1

The difference in estimating the length of walking roads according to the existing methodology and according to the proposed one is 6.6% and 4.9%. In taxation, such a difference is considered acceptable

During the field period, a ground survey of the road and trail network is carried out. During the examination, the road-tropine network diagram shall be checked and corrected, if necessary, based on the decryption data. At the same time, this scheme includes paths not identified on aerial photographs, for which purpose points are taken on the track of the unknown trail after 30 meters, the location of it on the aerial photograph or space photograph is located by linking it to the elements of the visible area on the picture. The found points are applied to the snapshot and connected to each other, forming the path of the trail.

To assess the quality of paths and roads, their width and type of coverage are determined by the elements. As the element of the path network means a section of the path from one intersection of the path with another path or road to the next intersection. In addition, a visual assessment of the elements of the walking path network is made according to 3 scales: on the scale for assessing the aesthetics of the surrounding forest park landscape paths (according to N.M. Tyulpanov) [2], on the scale for assessing the load of the path network [1; 3] and on the scale for assessing the patency of the path network [1; 3].

At the second stage, the obtained data on the area of the road-path network are compared with the normative. If according to the inventory data the area of the road-path network exceeds the normative, then part of the paths is not included in the designed road-path network, in that way, work on their improvement will not be carried out on these paths. First of all, are excluded the paths passing through the territory with stages fourth and fifth of recreational digression of the forest (the areas most affected by the recreational load), and the trails with the lowest scores on the scales. At the same time, it is necessary to ensure that the exclusion of elements of the path network does not disconnect the other elements. All roads are included in the design of the road-path network. The exclusion of the elements of the path network is made until the area of the road and path network becomes normative.

In the event that the area of the existing walking road and path network is less than normative, it is recommended to initially design objects of mass visits (beaches, recreation areas, etc.), and only then design new roads to it.

When landscape taxation of forest-park forests is assessed, such a landscape-architectural indicator as recreational digression of the forest. To increase accuracy and reduce the cost of landscape taxation of this landscape-architectural indicator, a technique was developed for its interpretation by aerospace photographs. Deciphering signs of recreational digression of the forest are shown in Table 3.

Table 3. Decoding signs of the stages of recreational digression of the fores.

Decryption features	Stages of recreational forest digression			
	I	2	3	4
1. Location of the surveyed area	in depth the wood	intermediate	near objects of mass recreation	
2. Road and Trail Network:				
-area	—	< 5 %	< 8 %	< 15 %
- location density	—	rare	intermediate	branched
- assessment of the load of paths	I point	1-2 point	3 point	4 point
-assessment of cross-country path	1 point	1-2 point	2 -3 point	3 point
3. Series forest park landscap	1б	1б, 1a	1a, 2б	2a

4. Placement of small architectural forms	not	not	there is	there is
5. Undereatage and undergrowth	uniform		intermediate	group

The difference in the assessment of recreational forest digression according to the existing methodology and according to the proposed one is given in Table 4.

Table 4. Comparison of the interpretation data of the stages of recreational digression of the forest with data obtained by the method of N. S. Kazanskaya [3].

Stage of recreational forest digression	Area, ha, defined		Distinction ($P_1 - P_2$)	
	by decryption (P_1)	forestry methodology (P_2)	ha	% to P_2
1	-	-	-	-
2	769,39	778,6	-9,21	-1,18
3	127,95	119,3	+8,65	+7,55
4	0,56	-	+0,56	-

The difference in the assessment of the areas of recreational forest digression determined by decoding aerospace photographs does not exceed 7.55%. In landscape taxation, when assessing indicators using the eye method, such a difference is considered acceptable.

Currently, landscape-architectural indicators in landscape taxation are evaluated by ocular method [2; 3]. The use of remote methods will increase the information content and accuracy of the assessment of forest parks. For this, a technique using ground digital images is proposed

During the on-site inspection of the territory in accordance with the existing methodology, three landscape-architectural indicators were assessed: landscape aesthetics, sanitary-hygienic assessment and patency of the surveyed forest-taxation site.

Taxation was carried out using the ocular method. To assess the landscape-architectural indicators, the scales used in landscape taxation were used [2; 3].

Simultaneously with the assessment of landscape-architectural indicators, a ground-based survey of the examined sections with a digital camera was carried out.

In order to know in office conditions which section is depicted in the image, information was recorded about each examined section. The shooting location was chosen so as not to shoot against the sun. Digital images on 58 sites were taken.

In the cameral period, by studying the images of the surveyed areas on a computer monitor, landscape-architectural indicators were evaluated.

The difference in the assessment of landscape and architectural indicators by the existing methodology and by the proposed one is given in the table 5.

Table 5. The accuracy of the assessment of landscape and architectural indicators.

Method of taxation		Quantity of areas, pieces								
		aesthetic class of forest park landscape			passability of area			sanitary and hygienic assessment		
		1	2	3	good	average	high	high	average	weak
Existing		24	25	9	14	21	23	13	24	21
Proposed		25	25	8	16	19	23	11	26	21
Difference	pieces	1	0	1	2	2	0	2	2	0
	%	7	0	11	14	18	0	15	14	0

An analysis of the data given in the table showed that the maximum deviation in the assessment of landscape-architectural indicators is 18% (when assessing the average cross-country ability). The average deviation in assessing the aesthetics class of the forest-landscape landscape is 6.0%, in assessing the patency of the site - 10.6%, and in the sanitary-hygienic assessment - 9.6% [4]. In taxation, when evaluating indicators using the eye method, such a difference is considered acceptable.

In addition, when designing events in forest park forests, it is obvious that a digital image of the highlight will be more informative for the designer than, for example, assessing its aesthetics on a three-point scale.

In order to estimate such a landscape and architectural indicator as the closure of the tree bed, images obtained during the shooting of the floor from the ground were used, the optical axis of the camera was directed vertically upwards. Then, using the electronic planimeter, the area of the crowns was calculated and the canopy closeness was calculated. Simultaneously with the shooting, the canopy closeness was assessed visually. Comparison of the data on the canopy closeness obtained by eye and according to the proposed methodology showed that the difference in average values is insignificant and amounts to 0.08 units. In landscape taxation, when assessing landscape-architectural indicators using the eye method, such a difference is considered acceptable.

3 Conclusion

The proposed methods for designing a walking road-path network and assessing recreational forest digression using aerial photographs will improve the quality of landscape taxation.

Assessment of landscape-architectural indicators: aesthetics of the landscape, sanitary-hygienic assessment, patency of the examined forest taxation area and closeness of the canopy of the forest stand, made according to the proposed methodology, based on the interpretation of digital terrestrial photographs using a computer, increases their information content and reliability.

References

1. O.S.Artemjev, *Methods of taxation of urban plantings* (SibGTU, Krasnoyarsk, 2003)
2. N.M.Tulipov. *Forest park economy: textbook. allowance* (Stroyizdat, L., 1975)
3. O.S.Artemjev, R.N.Matveeva, O.F.Butorova. [and others.] *Forestry basics in forest parks* (VNIItslesresurs, M, 1999)
4. O.S.Artemjev Forest J. **4,** 137-139 (2011)

How to apply remote sensing techniques to precision farming: 2 practical cases by Garnata drone

Javier Rodríguez Utrilla [1,*], *Álvaro* Nogués Palenzuela [1], *Javier* Cabello [2], *Domingo* Alcaraz Segura [3] and *Emilio* Guirado Hernández [4]

[1] Agroforestry responsible in Garnata Drone SL. 18013 Granada, Spain
[2] Department Biology and Geology. Andalusian Center for Assessment and Monitoring of Global Change, 04120, Almeria, Spain
[3] Department of Botany, University of Granada, 18071, Granada, Spain
[4] Soft Computing and Intelligent Information Systems research group, University of Granada, 18071 Granada, Spain

Abstract. In this report we show 2 different cases in which Remote Sensing can help Precision Farming techniques to optimize and improve profits in the agricultural sector. First, we`ll show a practical case in an olive trees field and second an example of how to monitor a greenhouse crop with aerial images and deep learning techniques.

1 Introduction

Terms like precision agriculture (PA), precision farming, site-specific crop management or even site-specific farming suggest that agricultural management can be practised with high precision. Consequently, PA provides an alternative and realistic means to reduce and optimize the use of potentially harmful compounds and thus can promote a healthier environment for humans [1].

The world's food scenery is changing fast rising the global demand and increasing the cost of agricultural inputs [2]. The adoption of this kind of technology is lower than 5% usually in countries, therefore is more than 30% in USA [3].

The general steps of PA practice are data collection, field variability mapping, decision making, and finally management practice. In particular, it is critical to obtain up-to-date images/maps during the process of decision making, thus field variability could be mapped using remotely sensed imagery. Drones play a special role in this field [4].

Hence, it is critical that farmers/Governments know where these variations exist in their fields and so they can adjust their farming practices accordingly [5]. Specifically, changes in remotely sensed reflectance can be detected before symptoms such as infections become visible to the human eye [6]. Because of that, we can advise our clients and calculate terms of productivity, fertilization maps, irrigation studies, early detection of pest/diseases [7], tree counting and damage assessment or automatic detection of areas/species of high

* Corresponding author: jrodriguez@garnatadrone.com

ecological value for the design of management plans. All the factors that we consider are those that we obtain from the multispectral index generation, as NDVI, NDWI, RVI, RI, GNDVI [9], and we analyze soil and plant, climate and DTM.

Here we present 2 different cases of study to show how this methodology can be adapted to any kind of crop, it can be useful in greenhouse crops too, improving and modernizing the agroforestry sector using the information obtained through remote sensing and changing the traditional farming practices.

2 Material and Methods

2.1 Study Sites

Two different field trials in Andalusia are included in this report. The first trial (Case A) is an Olive Trees field located in a production farm in the region of "Montes Orientals" in Granada province, Spain (37°23'33.93"N, 3°24'40.93"O) (Fig. 1).

Fig. 1. Study Area and location of olive trees field. Image Source: Google Earth.

The area has 11000 olive trees in 51 hectares of *Picual* variety, planted in 2006 at 7 m x 6m with drip irrigation system incorporated.

The second trial (CASE B) are three greenhouses situated in El Ejido, Almeria province (36°44'14.68"N, 2°46'46.21"O), Spain (Fig. 2). First and second greenhouse with chocolate pepper and last of them cucumber, three in early stages.

Fig. 2. Study site for case II. Three green houses in El Ejido, Almeria, Spain. Image Source: ArcGIS.

2.2 Image Acquisition and Processing. Analysis overv

The airborne campaigns were conducted using an UAV Parrot Disco AG Pro with a multispectral camera Parrot Sequoia with 4 different channels, Red, Green, Near Infrared and Red Edge, operated by Garnata Drone SL. The flights were programmed with Pix4D capture software, according to several factors, including the orography, cultivation system, plantation frame, etc.

CASE A) Olive Trees. We plan 2 flights per year, first before Spring (February 2018). We decided to carry out this test prior blooming to evaluate the real status in the crop field and design the amendments which the farmer needs to apply for in order to correct the worst areas. Second flight in Summer (June 2018) during the drought and before the harvest. In this flight we can evaluate the historical status in the olive tree, in case it's necessary to correct the irrigation plan. Finally, also we can check the effect that our recommendations to correct that.

The height was 120 m and overlap selected was more than 85% for a ground sample distance (GSD) of 13 cm/pixel.

CASE B) Green House. We plan a series of multispectral flights with a height of 80 m an overlap more than 90% that give us a GSD of 7,13 cm/pixel. The temporary distance between flights is one month.

With images obtained from the drone, in both cases, we generate ortoimaging processed by Pix4D software (Ecublens, Switzerland) and spectral index maps as NDVI (1), RVI (2), GNDVI (3) and NDWI (4) using SIG software, that allow us to detect affected areas that show high variability in one of those indexes, so we can separate the field into sectors based on the data found, using Object Based Image Analysis (OBIA) and Deep Learning.

$$NDVI=\frac{NIR-Red}{NIR+Red} \quad (1)$$

where, NIR – Near Infrared, and Red – Red Channel

$$RVI=\frac{NIR}{Red} \quad (2)$$

$$GNDVI=\frac{NIR-Green}{NIR+Green} \quad (3)$$

where, Green – Green Channel

$$NDWI=\frac{Green-NIR}{Gren+NIR} \quad (3)$$

2.3 Field Measurements

CASE A) Olive Trees

Once we know the "Problem zones", we can sectorize the field and take samples in areas with low spectral index values to create a complete study of vegetable variables from lab data (soil, water, leaf analysis, fat yield of the olive, etc...) which will explain the causes of anomalies and allow us to treat them in the proper way to optimize the crop productivity.

We received simultaneously the historical information and data of the crop field and produced an examination of the plantation that details all the necessary information for the decision-making process.

CASE B) Green House

We take Leaf Area Index (LAI) samples in several random points inside the green house (Fig. 3). Every point is measured by submetrical GPS. This measurement must be done at same time we fly UAV, so we can compare spectral image and ground control points/

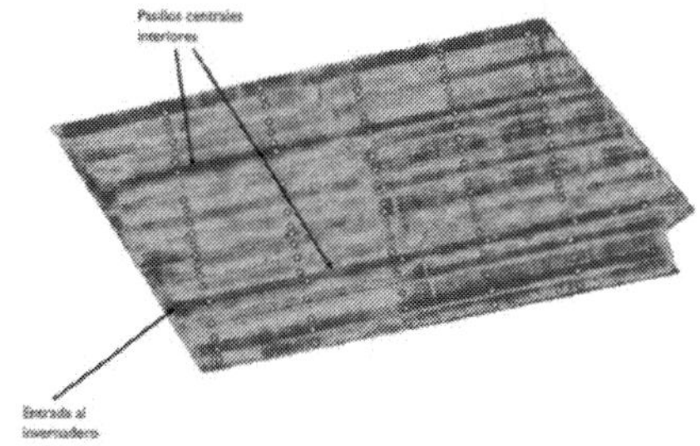

Fig. 3. Yellow points represent measurements of LAI taken by technicians at the same time UAV is flying. Image Source: Andalusian Center for evaluation and assessment of global change – CAESCG, Almeria, Spain.

3 Results and discussion

3.1 Olive Fields

Using Deep Learning, we generate a classification that allows us to create a database in which each tree is related to the values of the vegetable variables taken into account.

We then recommended amendments to the property owner based on the sectorized maps (Fig. 4) who combined information from spectral index and plant soil samples taken in the problem areas.

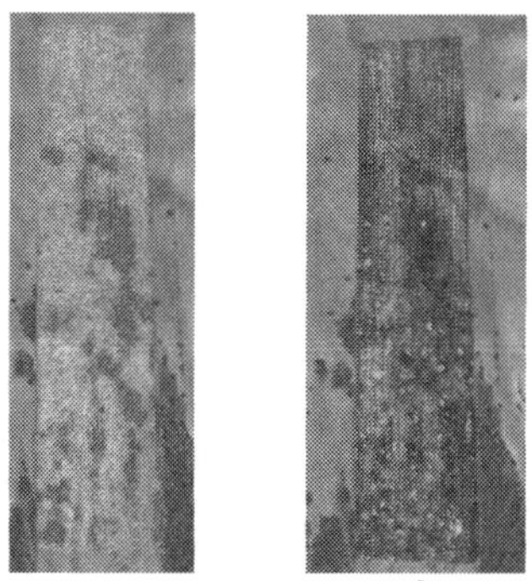

a *b*

Fig. 4. Diagnosis map. a) Left February 2018 flight. In red worst areas that needs to be correct. b) Right February 2019 flight, the red zones have decreased after our amendments recommended. Image Source: Garnata Drone SL.

Below in the Table 1 is a resume of last three years with and without assessment.

The average number of harvest per tree and hectare			
Year	**Kg per tree**	**Kg per Hectare**	**Advice Service**
2017	18.18	3995.60	No
2018	7.27	1597.80	Yes (Start)
2019	**22.72**	**4993.40**	Yes
The average of profit per tree and oil per each kg of olive			
Year	**Oil per each Kg (%)**	**Profit per tree (€)**	**Advice Service**
2017	22	12.72	No
2018	26	7.18	Yes (Start)
2019	**20**	**14.19**	Yes

Table 1. Resume values since 2017 with and without assessment.

3.2 Green House

The spectral data taken by drones will be validated with sensors arranged inside the greenhouses and treated by techniques of Object-Oriented Image Analysis (OBIA) and Deep Learning.

This will allow us to identify regions (Fig. 5) of the crop that present anomalies in the value of the greenery of the vegetation. These anomalies are directly related to the elements that can damage the crops. The results will be transferred to the technicians of the company, so that they can be applied to the treatment and monitoring of the status of the crops.

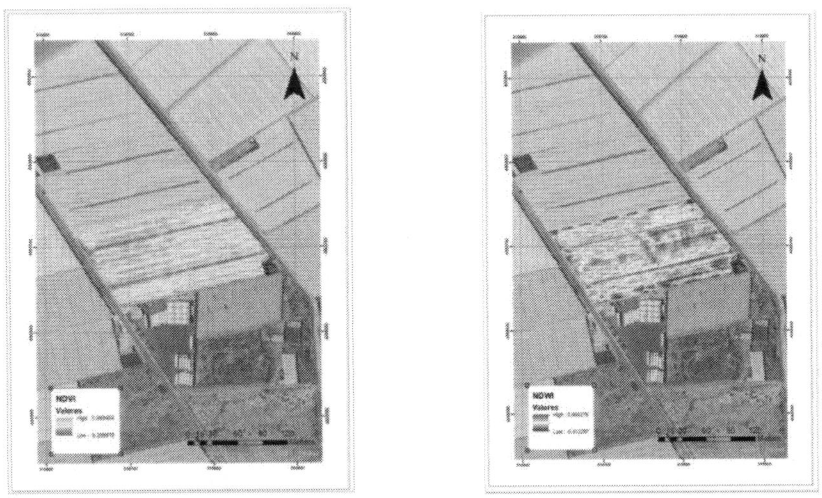

NDVI **NDWI**

Fig. 5. NDVI index in greenhouses (Left) and NDWI index (Right). Clearly the zones were sectorized based on the values of spectral information.

This would allow us to obtain primordial information for the development of monitoring systems for the greenhouses' crops that would have high spatial and temporal precision.

4 Conclusion

The monitoring of crops through spectral images is a reality in the rural world. These types of images, which provide information on functional aspects of the plants, have traditionally been obtained from sensors on satellites. However, spectral images from drones are equally effective and much more affordable.

Applying this methodology either in the open air or in greenhouses allows us to improve the physiological state of the crop by perfecting the application process to different plants and crops of necessary treatments, therefore improving productivity while simultaneously taking part in helping to protect the environment minimizing the inputs of phytosanitary and fertilizers, as we achieve better matching of fertilizer applications to crop yield.

Benefits to society include creation of high technology jobs and mitigation of environmental pollution arising from over – application of nitrogen and other fertilizers.

Therefore, is crucial that producers and scientists work together to create the future agriculture. This method does not replace all other sources of information, but help to improve it.

References

1. Zhang, Chunhua & Kovacs, John. The application of small unmanned aerial systems for precision agriculture // A review. Precision Agriculture. 2012. 13. 10.1007/s11119-012-9274-5.

2. R. Kent Crookston. A Top 10 List of Developments and Issues Impacting Crop Management and Ecology During the Past 50 Years. 2016.

3. Santhosh K. Seelan, Soizik Laguette, Grant M. Casady, George A. Seielstad. Remote sensing applications for precision agriculture // A learning community approach. 2003.

4. Swain, K. C., Thomson, S. J., & Jayasuriya, H. P. W. Adoption of an unmanned helicopter for lowaltitude remote sensing to estimate yield and total biomass of a rice crop // Transactions of the ASABE, 2010. 53, 21–27

5. Malthus, T. J., & Maderia, A. C. High resolution spectroradiometry: Spectral reflectance of field bean leaves infected by Botrytis fabae // Remote Sensing of Environment. 1993. 45, 107–116.

6. Chunhua Zhang, John M. Kovacs, The application of small unmanned aerial systems for precision agriculture // A review of Precision Agriculture. 2012. Volume 13, Number 6, Page 693.

7. Anastasiia Safonova, Alexey Rubtsov, Siham Tabik and Domingo Alcaraz-Segura. Application of convolutional neural networks in detection of forest damage caused by the polygraphus proximus beetle // Remote sensing. 2019, 11(6), 643.

Navigation network construction method for survey points using cartographic urban environment information

Ignat N. Penshin[1,*], *Larisa G.* Evstratova[1]

[1]State University of Land Use Planning, 105064, 15 Kazakova str, Moscow, Russian Federation

Abstract. This research solves the problem of navigation network construction for panoramic survey points inside courtyards without using road network information. The described method of navigation network construction takes into account a number of geometric conditions that detecting and identifying the regularities of the decisive routing rules construction. An algorithm for constructing links between panoramic shooting points, based on the surrounding area and linear objects territory information, is presented. An intelligent agent of geographic information system has been developed that executes this task based on the proposed decision rules. The goals and objectives of the research are presented, input data array to solve the problem is considered, the formulation of geometric conditions, structure and object implementation of the algorithm are set out in this work. The task is viewed in terms of the perspective of graph theory, implemented in the DBMS PostgreSQL and illustrated in the QGIS geographic information system.

1 Goals, objectives and subject of the research

The active development of digital modeling of the urban environment poses new challenges to the developers of intelligent agents of geographic information systems. Currently, there is a sufficient abundance and variety of freely distributed spatial data arrays that display relevant environmental information. In this situation the construction of unique analytical models and derived data arrays that promote the consumer becomes the most demanded product.

The purpose of this research is to build a navigation network for a web service for viewing street panoramas. A distinctive feature of this service is the emphasis on intra-quarterly shooting. This condition forms the task of building navigation within courtyards. A study of the quality and quantity of cartographic data on the urban road network showed a significant decrease useful information about roads inside courtyards. This conclusion is confirmed by the principle of classification of the road network, approved by the global Open Street Map (OSM) community, which prioritizes the road in the road network according to its connecting properties. Natural trails, footpaths, bicycle paths, and even

* Corresponding author: ignat.penshin51@gmail.com

asphalt roads inside courtyards, due to the low priority of logistic importance, may have an unknown status or may not exist at all due to the lack of inspection or the inability to determine the relationship with the nearest roads in cameral conditions. In this situation, even with direct visibility between neighboring points of photographing, creating navigation between panoramic images based on road networks becomes difficult or impossible. Therefore, the classical method of using the road network to navigate through panoramic surveys becomes irrelevant; under these conditions, the developed intelligent agent solves this problem.

Further disclosure of the topic requires an unambiguous definition of the following concepts:

A spatial routing network is an object implementation of a graph in the form of a set of linear and point facilities. Adjacency and incidence of objects is expressed by the storage of attribute information by a point or linear network object.

Panoramic survey points - photographic points, the position information of which was obtained using a GNSS receiver. They are considered as graph vertices and are point objects with spatial coordinates on an ellipsoid (LBH).

The neighborhood of a given vertex is the set of all vertixes of the graph adjacent to this vertex through the incident edge.

A private graph is an initially undirected graph constructed for one initial vertex. A private graph defines a neighborhood of a given initial vertex. After constructing a private graph the edges can be represented as oriented. The concept was introduced to solve the tasks of the research.

To start the process of creating a spatial routing for an array of points without reference to the road network, it is necessary to determine the initial data and set a task for constructing decision rules [1-6]. A solution is proposed and implemented in which an array of points and the connections between them are considered as an isomorphic undirected graph. The vertices of the graph correspond to a layer with point objects, and the edges correspond to a layer with linear objects.

Non-orientation indicates the presence of direct and reverse directions of movement between any pair of adjacent vertices. Each layer has its own table in the GIS Database. Each point feature has its own id, and the table for the line feature layer contains columns of attribute information about the id of the start and end points of each line. Therefore, we have complete information about the incidence of all elements of the graph with each other. The formation of the graph occurs as a result of determining the neighborhood for each vertex, therefore we can say that all private graphs with the beginning at the points of the array define each other and are uniquely connected - it follows that the property of isomorphism is inherent in the graph.

2 Input data array

Spatial routing is based on a number of geometric conditions, as well as conditions for interaction with surrounding objects. An intelligent agent must interact with an array of spatial data to fulfill decision rules [7]. Import of actual spatial data was made from the Open Street Map (OSM) server. The content and structure of the input data array for GIS intelligent agent located in the DBMS tables and presented in Table 1 and Figure 1.

Table 1. The content and structure of the DBMS tables of the input data array.

Table name	Content	Table fields and data types
	An array of panoramic	1. ID (integer) – the number of point; 2. Geom (geometry) – spatial coordinates of a

Panoramas	survey points from field surveys	given point; 3. Region (integer) – affiliation of the point to the survey area;
Buildings	An array of area objects with the building tag downloaded from the OSM server	1. ID (integer) – the number of area object; 2. Geom (geometry) – spatial coordinates of a given area object;
Highways	An array of line objects with the highway tag downloaded from the OSM server	1. ID (integer) – the number of line object; 2. Geom (geometry) – spatial coordinates of a given line object;
Edges	An array of linear objects that connects the panoramic survey points from the Panoramas table. The table is empty before the intelligent agent starts	1. ID (integer) – the number of a given line object (edge); 2. Pano_1 (integer) – Panoramas.ID number of the starting point of line object; 3. Pano_2 (integer) – Panoramas.ID number of the ending point of line object; 4. Geom (geometry) – spatial coordinates of a given line object;

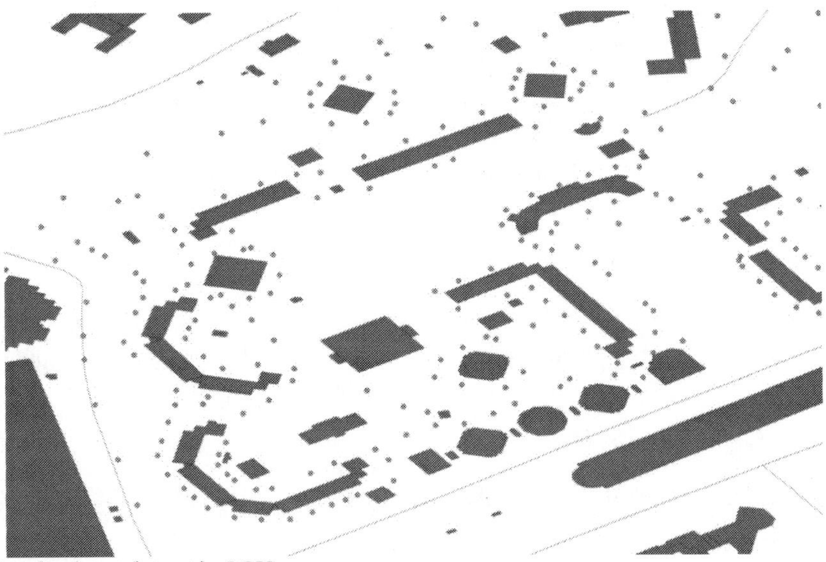

Fig. 1. Mapping input dataset in QGIS.

3 Geometric conditions

Consider the sequence of geometric conditions for building a network.

1. The maximum distance between points (dist_max) does not exceed 100 meters. This restriction is due to the resolution of photographic systems and provides a physical access to observe an adjacent point of photography, while viewing this panoramic image. The distance value is a variable and is an input parameter for the intelligent agent.
2. The line (edge) connecting adjacent points of photography must not intersect any areal object with the tag 'building'. This condition ensures direct visibility between adjacent points of photography.

3. The line (edge) connecting adjacent points of photography must not intersect any linear object with the tag 'highway'. This condition ensures the navigation zone within one city block, and also limits the possibility of building incorrect connections with the intersection of the roadway and violating the principle of logicality of spatial routing.

4. For each partial graph with an starting point N and neighborhood points satisfying the above conditions and belonging to an array of points in the table 'Panoramas', a necessary condition is the minimum central angle between any pair of edges (centr_ang_min) which is incident to the starting point N and to the corresponding points of the neighborhood. This fact allows to avoid cases in which the edges of a private graph are close to co-directed - it violates the principle of logicality and intuition of navigation. The executed studies allow us to conclude that a rectangular network is an ideal option. Having actual cartographic data, an approximate minimum central angle value of 40°-60° was obtained experimentally (the absolute value of the azimuth difference of any pair of edges of a private graph with a starting point N). The value of the central angle is a variable, as well as the input parameter of the intelligent agent.

5. A private graph must not have more than 4 incident edges. This restriction is due to the convenience and intuition of user navigation and corresponds to the authors' idea of an ideal navigation network similar to a rectangular network. This restriction applies to any private graph.

6. If there are an excessive number of neighborhood points for the starting point N of any private graph, the incident edges should be sorted by increasing their length (geometry) and a restriction on the use of the first four edges from the resulting list should be established. Thus, the four most suitable points of the neighborhood will be determined.

7. If the value of the central angle between any pair of edges of the private graph which are incident to the starting point N and the corresponding points of the neighborhood is less than the angle in the set variable - then the priority is given to the edge with the shorter length (geometry), and the edge with the longer length (geometry) is removed from the private graph. In this case, the starting point N and the neighborhood point which is incident to the remote edge cease to be adjacent.

8. The navigation network should provide the ability to move the user in the forward and reverse direction between any pair of incident points of panoramic shooting. In this case, the navigation network should be considered as a graph with non-oriented (undirected) edges, and any incident points should be connected by lines in the forward and reverse directions.

4 Building decision rules and their implementation by an intelligent GIS agent

The developed intelligent agent is based on the implementation of crucial rules:

1. For each point of panoramic shooting (table Panoramas) we construct a buffer zone with a radius equal to the variable of maximum distance between points (buff_radius = dist_max):

1.1. To all neighborhood points (table Panoramas), crossed by the buffer zone of this point, we build linear objects - we get a private oriented graph for this point.

1.2. Delete linear objects intersecting linear (table Highways) or areal objects (table Buildings) that have fallen into the buffer of this point. Neighborhood points satisfying 1, 2, and 3 geometric conditions are obtained.

2. For each private oriented graph, define all central angles (centr_ang) between all linear objects (edges).

3. In each particular oriented graph, we make a selection of pairs of linear objects (edges), the central angle between which is less than the set minimum value (centr_ang <centr_ang_min).

3.1. An object with a longer length is deleted from two conflicting. Thus, geometric conditions 4 and 7 are satisfied.

An example of selecting edges for points in a neighborhood of a private graph according to decision rules 1-3 is shown in Fig. 2.

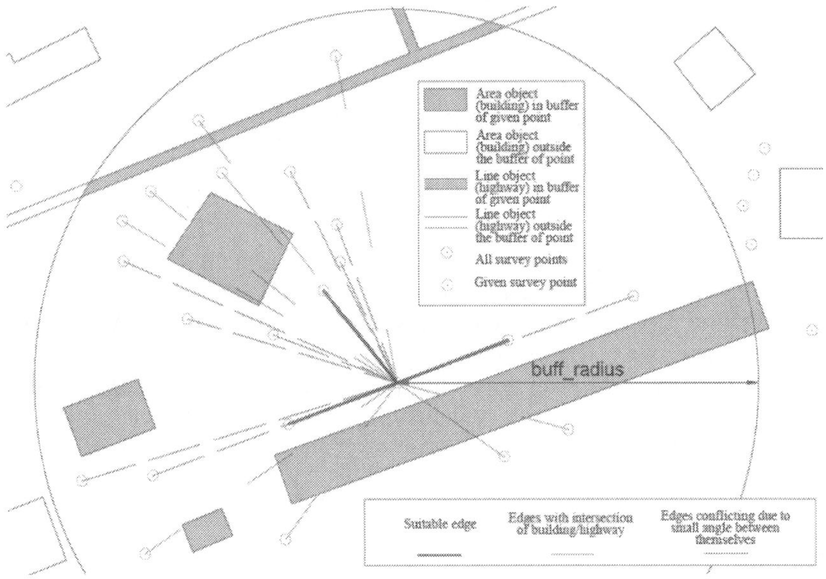

Fig. 2. An example of selecting edges for points in a neighborhood of a private graph according to decision rules 1-3.

4. Sort the edges in private graphs in increasing order and save the first 4 elements. Therefore, geometric conditions 5 and 6 are satisfied.

5. Complete the network with duplicating edges in the opposite direction. We obtain an undirected isomorphic graph according to geometric condition 8.

Since private graphs were formed independently of each other, the isomorphism obtained from paragraph 5 leads to a violation of geometric conditions 4 and 5.

6. Repeat paragraphs 3 and 3.1 for undirected private graphs with the removal of linear facilities (edges) in the forward and reverse directions.

It is necessary to filter linear facilities that divide the resulting rectangular network into groups of triangles.

7. In pairs of linear objects forming an acute angle at one common vertex and connected by another linear object through the remaining vertices, an linear object with a longer length is deleted.

A graphic illustration of the steps involved in implementing decision rules by a GIS intelligent agent is shown in Fig. 3.

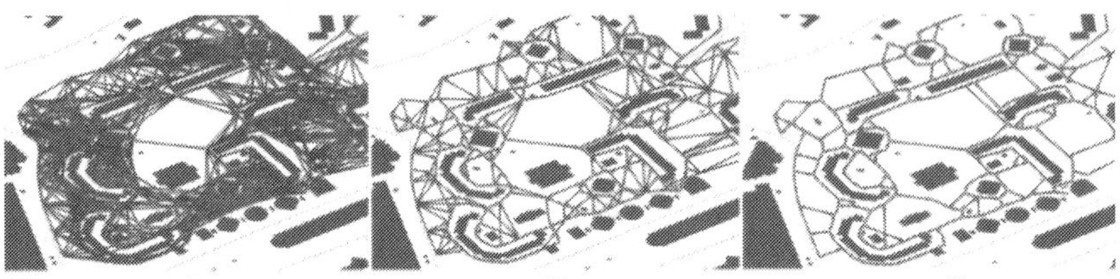

Fig. 3. Steps for implementing decision rules by an intelligent agent:
Stage A - Implementation of 1 decision rule;
Stage B - Implementation of decision rules 1-5 with the initial reduction of the graph to an isomorphic form;
Stage C - Implementation of decision rules 1-7.

Conclusion

The proposed method for constructing a spatial routing network provides movement within courtyards based on the construction of decision rules for a number of developed geometric conditions and network requirements. The research results allowed us to build a routing network based on real data for 1,500,000 survey points, which completely covers Moscow city. Thus, the developed GIS intelligent agent has the ability to quickly generate a high-quality set of rational solutions, and it can be used for operational, reliable spatial routing inside courtyards without using information about the road network.

References

1. Hand, D. J., Data Mining, *Encyclopedia of Environmetrics*, **2** (2006)
2. Brown, D. G., Riolo, R., Robinson, D. T., North, M., & Rand, W., Spatial process and data models: Toward integration of agent-based models and GIS, *Journal of Geographical Systems*, **7(1)**, 25-47 (2005)
3. Bell, M. G., Iida, Y., Transportation network analysis (1997)
4. Munkres, J., Topology (2014)
5. Fotheringham, S., Rogerson, P., Spatial analysis and GIS (2014)
6. Lee, J., A spatial access-oriented implementation of a 3-D GIS topological data model for urban entities, *GeoInformatica*, **8(3)**, 237-264 (2004)
7. Kwan, M. P., & Hong, X. D, Network-based constraints-oriented choice set formation using GIS, *Geographical Systems*, **5**, 148-153 (1998)

E3S Web of Conferences **149**, 02002 (2020) https://doi.org/10.1051/e3sconf/202014902002
RPERS 2019

Problems, methods and technologies of combined 3D model construction for underwater and above-water relief

Andrey Gachenko*, *Alexei* Hmelnov

SB RAS, Institute of System Dynamics and Control Theory, 134 Lermontov street, Irkutsk, Russia

Abstract. In the article we describe the methods for construction of underwater 3D relief (bathymetry) and combining it with the corresponding ground relief (terrain) when the bathymetry information is of much lower quality than that of terrain. We have developed an algorithm for combining the low quality underwater relief with the higher quality ground relief using Delaunay triangulations. The resulting combined 3D model of relief has no noticeable artifacts and can be used to solve various hydrological tasks, such as computation of the HPP reservoir water level in dependence to the water discharges of the hydroelectric power plant. We have tested the technology for building combined relief 3D model in several scientific projects intended to estimate the changes in the coastline under various water flow scenarios in the basins of Angara River and Lake Baikal.

1. Introduction

The river bed of the Angara River on its more than 100 km stretch from the dam of the Irkutsk Hydroelectric Power Station (IHPS) to the start of the Bratsk Reservoir has rather complex structure with many islands, backwaters, channels and inflow junctions. The most dangerous from the point of view of potential flooding is the 67 km long area from the dam of IHPS to the end of the city of Angarsk. About 1 million people live here and the area contains the largest enterprises of the Irkutsk region. To consider the risks of flooding of urban areas caused by increased water discharges through the dam of IHPS and the other river Angara inflows due to extreme weather conditions it is necessary to build the relief models to determine potential zones of flooding. Regular space monitoring of the specified area may be used to measure the coastlines with the accuracy of several meters but it is a costly affair [1-9].

By using sufficiently accurate relief model it will be possible to estimate the dependence of potential flooding boundaries on the water discharges. Moreover, in conjunction with the period of extreme low water that began in 2014 in the basins of lake Baikal and river Angara the problem of estimating the boundaries of shallowing of the lower pool of the IHPS and the Bratsk reservoir is of great current interest.

* Corresponding author: gachenko@icc.ru

Thus, we need the river bed relief model. The model should take into account the shape of the river bottom. While the ground relief is represented by topographic vector electronic maps of various scales and various global DEMs (e.g. SRTM, ASTER), the information about the underwater relief of the Angara River is not available electronically. There exist some electronic bathymetry data for navigators, but they are fragmented and very imprecise. The only useful sources of the underwater relief information are the pilot charts, which are available in paper form only. To use the pilot chars we have to vectorize them (Fig. 1).

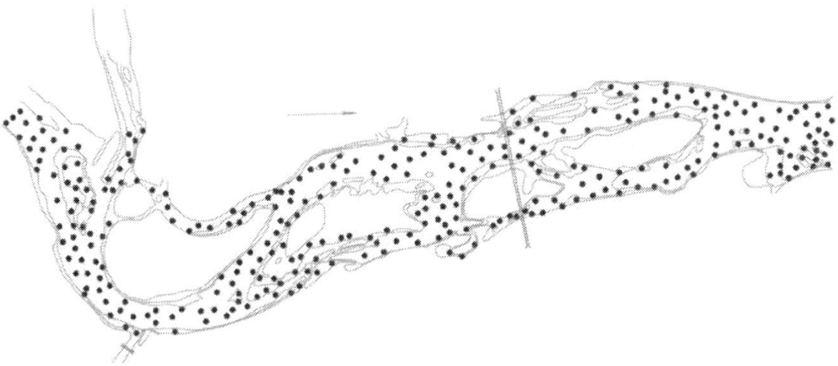

Fig. 1. The results of vectorization of the underwater relief of the Angara River.

2. Models and methods

To digitize the pilot charts we use the free version (7.99) of the tool Easy Trace. From the pilot charts we digitize the following objects: water depth contour lines, depth marks, some engineering structures and the coastline.

All the sheets of the pilot charts have arbitrary north direction, because they were rotated to place the sheet most effectively along the correspondent fragment of the river. Thus, after vectorizing we have to rotate and shift the sheets back to their positions on the topographic map. To find and perform the coordinate transformation we have developed a small utility, that takes several pairs of coordinates of some distinctive points on the both maps (it is required to specify at least two points), computes the transform parameters using the least squares method, and transforms the source map (pilot chart) to the coordinates of the target map.

After performing the coordinate transformation it turns out that the vectorized coastline strongly mismatch the more precise coastline of the topographic map and it is impossible to match these lines by just turning and shifting (Fig. 2).

To build a visually consistent digital terrain model using the water depth data of low quality we have developed an original approach, which uses digital map morphing (transformation) algorithms.

 https://doi.org/10.1051/e3sconf/202014902002

Fig. 2. Comparison of the coastlines on the pilot chart (blue) and on the topographic map.

The problem of matching map layers of different precision is quite common. It can also arise, for example, when trying to use some thematic map layers, which are based upon a topographic map of low scale, together with the map layers of higher scales. The resulting map will have evident artifacts such as contour lines, forest areas or roads in water. To improve the result we have developed a method of map morphing and software to perform the morphing. The main idea of the method for the underwater data is that first we should find a transform to match the coastlines. In fact we don't need the coastlines from the pilot chart themselves at all, because the topographic map version is much better and we can always use it. But we need the transform of the coordinate plane, that matches the coastlines: we will use the transform for the bathymetry layers to avoid intersection with the topographic map coastlines and get visually consistent results without noticeable artifacts.

To specify the contour matching parameters we have implemented a special task in our electronic map viewer program IrkGV (Fig. 3). Using the task operator sets correspondence between some distinctive points of the source (inaccurate) and target (more precise) maps. The correspondence may be set between individual points (green arrows, rarely used) as well as between contour parts (red arrows). The red arrows, which are adjacent on both contours, specify the match of the contour parts between them. And sometimes it is also required to suppress matching between some contour parts (for example, when one of them has some surplus details, which would make the resulting transform too complex) – to do it we use black arrows.

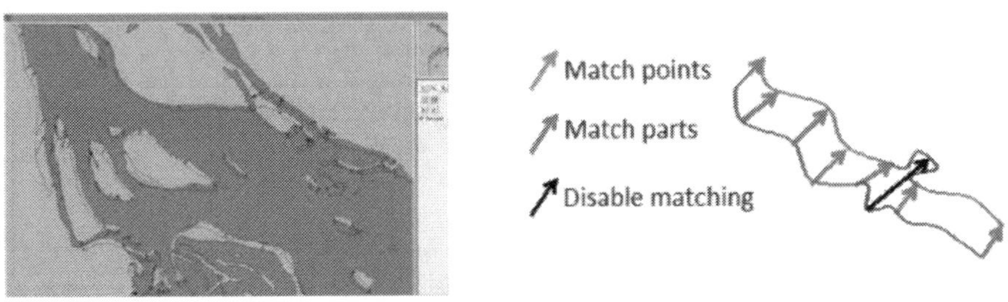

Fig. 3. The map morphing parameters task window. The arrows specify correspondences between individual points of the maps and between the contour parts.

The mapping of the points of the matched contour parts is performed using the linear transformation of the parametric coordinates of the curves. If the resulting transform will fail to match some distinctive points of the contours, it will be required to explicitly set correspondence between the points by red arrow and repeat the transformation. After setting all the matches between the maps we generate the displacement file which, besides from the explicitly set displacements (by the arrows) contains the displacements computed for the intermediate points of the matched contour parts.

To transform the source data we have developed a special tool using constrained Delaunay triangulation algorithms [10]. It takes the displacement file and builds a constrained Delaunay triangulation, which contains the displacement vectors in each of its vertices. The segments of the matched contour line parts of the source map are the constraints (hard edges) of the triangulation. For any point of the coordinate plane in the area of interest we can find the triangle, to which it belongs, and compute the displacement vector for the point using linear interpolation of the displacement vectors of the triangle vertices. So, the triangulation specifies transform of the coordinate plane, that we use for all the other low quality layers.

Finally we build the triangulation, which represents the combined relief. The map layers used for triangulation construction may play distinct roles: ground relief, water level marks, contours of coastlines, underwater relief. The underwater relief depth is measured from the water level, so we first use information about the water level marks to build an auxiliary triangulation, which is then used to determine the water level in each underwater data point.

While performing morphing for combining the underwater and ground reliefs we obtain a continuous transformation of the coordinate plane which can match the inaccurate contours of the coastlines with the more accurate ones. The transformation is then applied to the other layers of the underwater relief map (isobaths and depth marks). The map layers obtained as a result of the morphing are consistent with the information about the ground relief.

4. Results and discussion

We have applied the approach considered for the fragment of the Angara river on the 135 km stretch from the IHPS to the city of Svirsk. As a result we have obtained the digital 3D model of combined relief (Fig. 4). A digital terrain model of this kind was obtained for the first time and has no analogues.

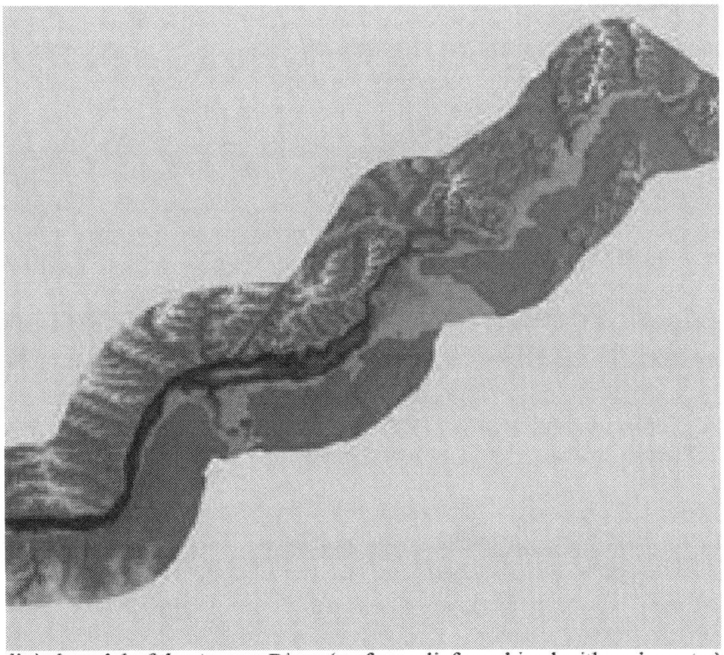

Fig. 4. Final digital model of the Angara River (surface relief combined with underwater) from IHPS to the Bratsk reservoir (Svirsk).

The same approach was used for selected regions of the Bratsk reservoir (Fig. 5,6). The Bratsk reservoir is extremely large, so we have created the combined relief 3D models for more than 50 areas of interest, but not for the whole reservoir.

Fig. 5. A fragment of the resulting map for the Bratsk reservoir near the Kamenka settlement. The bathymetry contour lines were obtained from pilot charts.

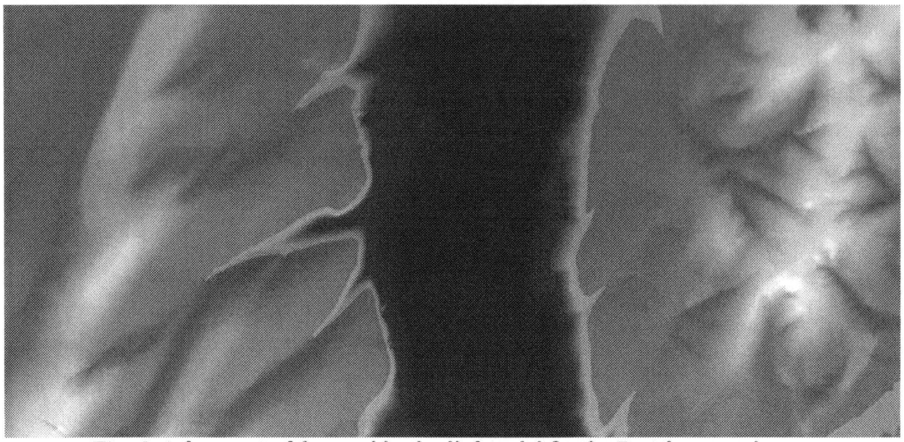

Fig. 6. A fragment of the combined relief model for the Bratsk reservoir.

We have also an experience of creating combined relief models without morphing, when all the data used have sufficient spatial precision. Thus, during implementation of a project for assessment of anthropogenic impact the digital terrain model of the area of Listvyanka village was created (Fig. 7) based on a high-resolution satellite image (WorldDem, 10 meters per pixel).

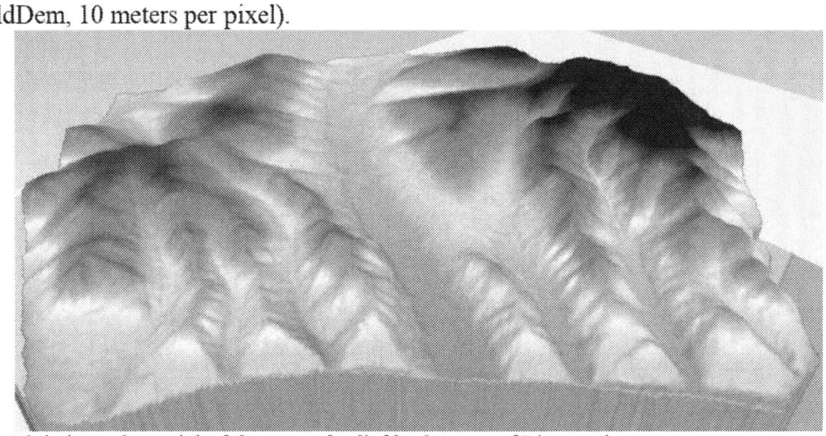

Fig. 7. Digital terrain model of the ground relief in the area of Listvyanka.

The underwater relief was obtained from the echo sounder measurements of the bottom relief near the edge zone of Listvennichny Bay. The resulting combined relef is shown on Fig. 8.

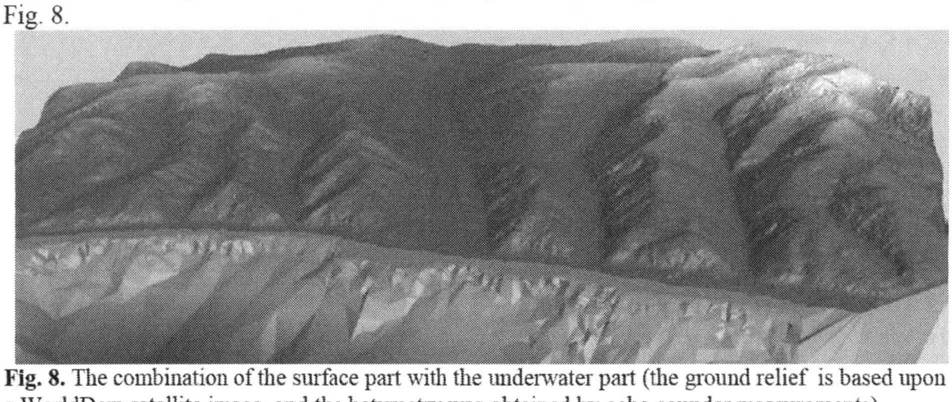

Fig. 8. The combination of the surface part with the underwater part (the ground relief is based upon a WorldDem satellite image, and the batymetry was obtained by echo sounder measurements).

Conclusion

The combined 3D models of ground and underwater relief make it is possible to solve various hydrological tasks for reservoirs, lakes and rivers. The technologies and methods described in the article have been tested on a number of projects and have shown their effectiveness.

The combined reliefs were used to assess changes in the coastline under various scenarios of water availability in the water basins of Angara and Lake Baikal. The created 3D models of relief were used as a part of hybrid geographic information system that allows you to simulate various scenarios of changes in the water level and to determine the zones of the coastline change.

This work was performed with partial financial support of the Russian Foundation for Basic Research (Grant No. 17-47-380005).

Reference

1. Grachev M A 2016 Maybe-Bag and Let it Be. Ecological crisis on Baikal: A mystery of the Century *Science first-hand* **68** pp 6–19.
2. Bychkov I V and Nikitin V M 2015 Control of level of Lake Baikal: problems and possible issues *Geography and environmental resources* **3** pp 5–16.
3. Bezrukov L A, Saveliev V A, Nikolskii A F and Podkovalnikov S V 1997 Baikal and hydropower: ecology and economy Geography and environmental resources 4 pp 158–168.
4. Atutova A A, Pronina N M and Tulohonova A K 1999 Hydropower and condition of an ecosystem of Lake Baikal (Novosibirsk: publishing house SB RAS) p 280.
5. Itskovich V B, Shigarova A M and Glyzina O. 2015 Change in the Hsp70 content in the Baikal endemic sponge Lubomirskia Baicalensis during decolorization and under hypothermia *Aktual'nye problemy nauki Pribaikalya*. pp 135-138.
6. Vashchenko B 2015 Aliens in Baikal *National Geographic Russia* **146** p 6.
7.Kravtsova L S, Izhboldina L A and Khanaev I V 2012 Disturbances of vertical zoning of green algae in the littoral area of the Listvennichnoye Bay, Lake Baikal, as a result of local anthropogenic effect *Doklad RAS* **2**. pp 227-229.
8. Kravtsova L S, Izhboldina L A and Khanaev I V 2014 Nearshore benthic blooms of filamentous green algae in Lake Baikal *J. Great Lakes Res*. **40** pp 441-448.
9. Abasov N V, Bolgov M V, Nikitin V M and Osipchuk E N 2017 *On regulation of the water-level mode of Lake Baikal Water Resources* **44** pp 407-416.
10.Skvortsov A V 2002 Delaunay triangulation and its applications (Tomsk: Tomsk University press) p 128.

Hyperspectral regression lossless compression algorithm of aerospace images

Assiya Sarinova[1,], *Alexander* Zamyatin[2]

[1]Toraighyrov named after Pavlodar state university, Electrical Engineering and Automation Department, 140000 Lomov street 64, Kazakhstan
[2]National research Tomsk state university, Institute of Applied Mathematics and Computer Science, 634050 Lenina avenu 36, Russia

Abstract. In this work, we propose an algorithm for compressing lossless hyperspectral aerospace images, which is characterized by the use of a channel-difference linear regression transformation, which significantly reduces the range of data changes and increases the degree of compression. The main idea of the proposed conversion is to form a set of pairs of correlated channels with the subsequent creation of the transformed blocks without losses using regression analysis. This analysis allows you to reduce the size of the channels of the aerospace image and convert them before compression. The transformation of the regressed channel is performed on the values of the constructed regression equation model. An important step is coding with the adapted Huffman algorithm. The obtained comparison results of the converted hyperspectral AI suggest the effectiveness of the stages of regression conversion and multi-threaded processing, showing good results in the calculation of compression algorithms.

Introduction

Hyperspectral aerospace images (AI) are images obtained from Earth remote sensing spacecraft (ERS), designed to solve problems in the field of applied research. Studies to compress hyperspectral AI are the most interesting; this is subject to a large number of publications. In hyperspectral AI for each pixel, the hyperspectral camera accepts light intensities for a large number of adjacent spectral ranges reaching several hundred. Hyperspectral AI due to the rich content of information are effectively used in the tasks of automated processing of remote sensing images. The stages of hyperspectral (AI) processing can include: internal representation of images, image conversion, geometric correction of scenes, and image preprocessing. Compression efficiency can consist of the following steps - a change in the internal representation of images, converting it to a compact form, and using pre-processing. Compression is the preprocessing and removal of redundancy in images. There are two types of redundancy in hyperspectral AI, spatial and spectral. These redundancies will allow the development of efficient compression algorithms.

Based on researches of hyperspectral AI in the field of compression [1–20] presented in the works of scientists from Russia, China, the USA, India, etc., it can be assumed that the developed methods and lossless compression algorithms for hyperspectral AI can be improved by reducing their computational efficiency and increasing the compression ratio by modifying the preprocessing steps using mathematical methods. In addition, new compression preprocessing steps can be proposed that effectively increase the compression ratio and reduce the compression process time. The latest best results in solving the compression problem, as in many other remote sensing problems, were obtained using well-known algorithms in their various combinations and groups, while having low compression ratios, the average compression ratio is 3.85.
Based on the analysis of lossless GI compression methods and algorithms, it should be concluded that the most effective ways to solve the compression problem are:
- taking into account spectral correlation, which gives certain advantages on the basis of the calculated correlation matrix;
– application of a new method of ordering GI channels;
– use of interpolation based on mathematical methods;
– arithmetic coding and the Huffman algorithm, which are the best among statistical methods;
– the use and organization of parallel compression processing to reduce the cost of computing resources.

In a previous study of the authors of a multi-stage compression algorithm [10], the main drawback remains - the computational efficiency of compression is low, large computational costs are required, which makes it difficult to use them to solve GI compression problems.

Therefore, in this work, a modification of the algorithm is proposed taking into account the correlation [10], in which an attempt is made to eliminate this drawback.

Modification of the algorithm with regression analysis. The essence of the modification is to calculate the linear regression coefficients between the values of the generating channel (master) and the regressed channels (slave, compressible) of the GI by forming arrays of differences between master and slave. When using a PC with different computing capabilities, the modification of the algorithm uses a different number of threads, corresponding to the number of available processor cores. The proposed modification of the lossless compression algorithm taking into account inter-band correlation and regression analysis will increase the compression ratio by more than two times in comparison with the use of universal archivers. The proposed algorithm for finding the best channel groups for a given correlation value will increase the efficiency of the channel subtraction (difference transformation) application.

1 Description of the approach to the problem of compression of hyperspectral AI

The sequence of processing steps and the compression algorithm:

1) calculation of the correlation between all pairs of AI channels and determining the sequence of channel coding and decoding;

2) regression conversion algorithm;

3) obtaining channel differences and their block conversion;

4) compression by a statistical algorithm.

We describe each step of the algorithm.

1 step. We calculate the values of the correlation matrix between all pairs of AI channels, revealing the most correlated groups of channel pairs. Based on the matrix, we

will form and determine the sequence of transformation (coding) and inverse transformation by the method of constructing a strongly branching tree.

2 step. Regression analysis based on step 1. Let us calculate the linear regression coefficients between the values of the generating and regressing channels of the hyperspectral AI by creating optimal values for generating arrays of differences between master and slave.

3 step. Compression by the Huffman statistical algorithm [2].

2 Description of the compression algorithm

2.1 Compression Sequencing

At the first stage, we transform the data structures based on the initial hyperspectral AI, using mathematical models for finding the correlation between all pairs of channels, and we determine the coding sequence. Knowing that hyperspectral AI are obtained in the spectrum of a single wave, we assume that a certain degree of dependence can be determined between pairs of channels. To determine the magnitude of this dependence, we use the formula for calculating the Pearson correlation coefficient (aka linear correlation coefficient).The formula requires two sequences of data, so we first extract two sequences of samples from a pair of channels. This happens as follows: we represent the two-dimensional data matrix of one channel in the form of a linear array (passing the matrix row by row from left to right and rows from top to bottom), then select a certain number of samples in it (we denote the number by the letter m) dividing the array into approximately equal segments. From the second channel, we extract a sequence of samples that are located in the matrix at the same positions as the samples from the first channel. We denote the obtained sequences by the letters x and y, and the unit values are x_i and y_i (i from 1 to m inclusive). It is also necessary to calculate the arithmetic mean values of both sequences of samples.

Counting the coefficients for all possible pairs of channels of the same image, we build the coding sequence - naturally, starting from larger values to smaller ones.

1 The Construction of the correlation matrix of AI channels.

2 Build a highly branching tree to determine the sequence of channel conversion depending on the correlation value using the Pearson correlation calculation formula.

3 Formation of the hyperspectral AI processing sequence for step 1.2, necessary for AI recovery.

Example. There is a 12-channel GI, respectively, the index tree will contain 11 vertices, we will build them in the correlation table, table. 1. From table 1 it is seen that the interchannel correlation increases, starting from channel 5. This suggests that the channels whose indices lie in the range [4-11] have a high correlation dependence, the maximum correlation value between channel 10 and 11.

Table 1. Correlation matrix of selected channels for compression

	0	1	2	3	4	5	6	7	8	9	10	11
0	1	0,35	0,29	0,3	0,23	0,23	0,22	0,23	0,22	0,23	0,22	0,23
1	0,35	1	0,34	0,34	0,3	0,27	0,29	0,29	0,28	0,29	0,29	0,3
2	0,29	0,34	1	0,6	0,56	0,51	0,53	0,52	0,5	0,5	0,49	0,49
3	0,3	0,34	0,6	1	0,89	0,85	0,85	0,85	0,82	0,81	0,8	079
4	0,23	0,3	0,56	0,89	*1*	*0,95*	*0,95*	*0,95*	*0,95*	*0,92*	*0,92*	*0,9*
5	0,23	0,27	0,51	0,81	*0,95*	*1*	*0,97*	*0,97*	*0,97*	*0,96*	*0,95*	*0,95*
6	0,22	0,29	0,53	0,85	*0,95*	*0,97*	*1*	*0,985*	*0,983*	*0,978*	*0,972*	*0,968*
7	0,234	0,29	0,52	0,845	*0,945*	*0,975*	*0,985*	*1*	*0,990*	*0,981*	*0,984*	*0,982*
8	0,23	0,29	0,50	0,82	*0,935*	*0,970*	*0,983*	*0,990*	*1*	*0,992*	*0,991*	*0,988*
9	0,23	0,29	0,50	0,81	*0,924*	*0,963*	*0,978*	*0,989*	*0,992*	*1*	*0,995*	*0,994*
10	0,224	0,295	0,491	0,803	*0,912*	*0,955*	*0,972*	*0,984*	*0,991*	*0,995*	*1*	*0,996*
11	0,23	0,30	0,499	0,798	*0,907*	*0,950*	*0,968*	*0,982*	*0,988*	*0,994*	*0,996*	*1*

3.1 Based on the constructed correlation matrix, we construct the L tree of indices - these are ordered pairs of channels (for example, channels numbered 10 and 11). Let there be 11 pairs and 11 vertices of the tree. Channel 12 under index [11] is converted through the first vertex of the tree [# 0] (the numbering of vertices and indices starts from zero), Channel 10 under index [9] is converted through the second vertex [# 1], Channel 9 under index [8] is converted through the third peak [# 2], etc., fig. 1.

3.2 Formation of a sequence of pairs:

Step 1. Find the maximum element among the available pairs in the correlation matrix, form the first pair of channels.

Step 2. We put in a set of ordered pairs (SOP) a new pair of channels.

Step 3. We remove from the table of the correlation matrix all pairs whose elements are already defined in the SOP.

Step 4. While all the elements are not in the SOP, we return to step 1.

#0	10 ->	11
#1	10 ->	9
#2	9 ->	8
#3	8 ->	7
#4	7 ->	6
#5	7 ->	5
#6	6 ->	4
#7	4 ->	3
#8	3 ->	2
#9	3 ->	1
#10	1 ->	0

a)

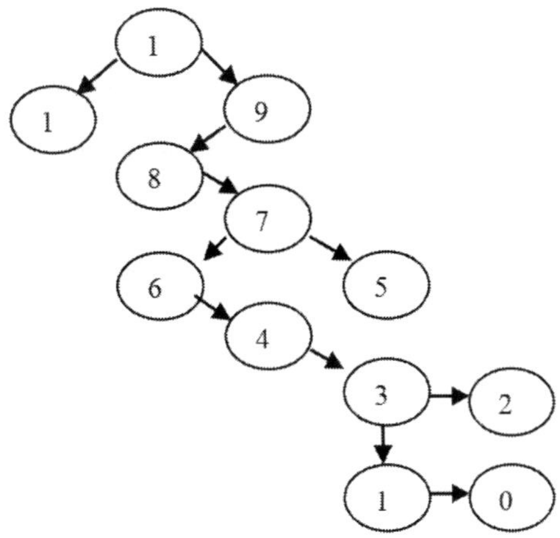

b)

Fig. 1. The tree of correlated channels:
a) the set of ordered pairs, b) the sequence of ordered pairs

2.2. Regression (regression) transformation

The essence of the transformation is to bring into conformity to the encoded pair of channels some structure that would:

1) allowed to unambiguously restore one of the source channels according to the data of another channel,

2) occupied as little disk space as possible.

The subtracted channel is converted using the linear regression equation (LRE) as an example. The generalized form of the transformation will be like this.

LRE - "Y = $a \times x$+b". Y are unknown values, x are the values of the main channel, y are the values of the second channel, we find a and b:

1) we find a - coefficient by intermediate calculations;

2) we find b - coefficient by subtracting the average of y and the available coefficient a;

3) calculate Y = $a \times x$+b. where x - values of the main channel, found LRE.

For example, the value of the generating channel x is equal to 918, y - the value of the second channel is 1312. Between the channels there is a weak correlation of 0.3636 (the difference between the channels x and y = 394), the coefficients a = 0.49, b - 784.01 are found. Then Y = 1233. The difference is $d = y - Y = 79$. The value is 79 <394 and requires fewer bits to store them.

We apply this type of generalization of LRE to hyperspectral AI. The coefficients of the constructed regression model $Y_i = a \times x + b$ for the generating channel were determined.

The idea of linear regression in our case is to find such real values a and b such that the matrix formed from the data of the encoded pair by the following formula: $d_{ij} = (x_{ij} \cdot a + b) - y_{ij}$, (hereinafter, x_{ij} and y_{ij} are the values in the matrices master and slave, respectively) would have in itself the smallest possible values. Below we will call the matrix d with the values d_{ij} the "difference matrix". A sufficiently good linear correlation index (close to unity), calculated previously, leads to rather low values of d_{ij}. Knowing the values of a, b and the matrix d, we can restore the matrix y from the values of the matrix x:

$$y_{ij} = (x_{ij} \cdot a + b) - d_{ij}$$

The standard formulas used to calculate linear regression parameters are:

$$a = \frac{\overline{xy} - \overline{x} \cdot \overline{y}}{\overline{x^2} - (\overline{x})^2}, \quad b = \frac{\overline{x^2} \cdot \overline{y} - \overline{x} \cdot \overline{xy}}{\overline{x^2} - (\overline{x})^2},$$

where:

$$\overline{x} = \frac{1}{i \cdot j} \sum_{i=1}^{m} \sum_{j=1}^{n} x_{ij}; \quad \overline{y} = \frac{1}{i \cdot j} \sum_{i=1}^{m} \sum_{j=1}^{n} y_{ij}; \quad \overline{xy} = \frac{1}{i \cdot j} \sum_{i=1}^{m} \sum_{j=1}^{n} (x_{ij} \cdot y_{ij}); \quad \overline{x^2} = \frac{1}{i \cdot j} \sum_{i=1}^{m} \sum_{j=1}^{n} x_{ij}^{\,2};$$

(m - image height, n - width).

In each slave channel we put the values a and b in the Double format (8 bytes, 15 decimal places in decimal format). Encoding the only main generating channel (generating the channel of the first pair) for the image, independently of the others, we can subsequently restore it first, then gradually restore all other compressed channels (as mentioned earlier, through the coefficients a and b).

The absence of losses during the conversion is ensured as follows. After receiving the matrix d and before writing it to the file, we round off the values to the nearest integer (in the case of a fractional part equal to 0.5, to a smaller integer). This does not prevent lossless recovery. Let's explain why. Let's pay attention to the formula: $d_{ij} = (x_{ij} \cdot a + b) - y_{ij}$.

The matrices x and y are integer; therefore, the quantity has the same fractional part (we denote it by q) as the number d_{ij} before rounding. We will denote the integer part of numbers by square brackets, the fractional by curly brackets. So q = $\{d_{ij}\}$ = $\{w_{ij}\}$. After rounding, rounded values are written to the difference file. We proceed to recovery:

$$y'_{ij} = (x_{ij} \cdot a + b) - d'_{ij} = (d_{ij} + y_{ij}) - d'_{ij}$$

$$y'_{ij} = (d_{ij} + y_{ij}) - d'_{ij} = q + y_{ij} - 1 = y_{ij} - (1 - q).$$

The value of w_{ij} remains unchanged, since a and b were previously calculated and recorded in the Double format (that is, without loss), the values of x are integer and do not undergo losses.

If q <= 0.5, то $d'_{ij} = [d_{ij}]$ and: $y'_{ij} = (d_{ij} + y_{ij}) - d'_{ij} = q + y_{ij}$.

So, rounding, we get the initial value y_{ij}.

If q > 0.5, то $d'_{ij} = [d_{ij}] + 1$, and: $y'_{ij} = (d_{ij} + y_{ij}) - d'_{ij} = q + y_{ij} - 1 = y_{ij} - (1 - q)$

The value (1 - q) <= 0.5, a y_{ij} is an integer. It also means: rounding, we get the initial value y_{ij}.

Thus, regression conversion does not incur losses. To reverse decode hyperspectral AI, we perform the following steps:

Step 1. Decoding arrays of differences by the Huffman algorithm.

Step 2. Formation of regression transformation arrays by finding the sums between the generating channel and its average value.

Step 3. Formation of the initial arrays based on the available coefficients of HRM and obtaining the initial data of hyperspectral AI.

3 Experimental researches

To determine the effectiveness of the proposed algorithm from the point of view of the degree of compression, as well as the limits of its applicability, a number of experiments were carried out on hyperspectral AI (*Aviris* remote sensing system), in table. 2.

Table 2. Hyperspectral AI Test Data

Count bands	Size	Size *(byte)*	*Count bands*	Size	Size *(byte)*
100	100*100	4080400	200	100*100	8160800
100	200*200	16160400	200	200*200	32320800
100	300*300	36240400	200	300*300	72480800
100	400*400	64320400	200	400*400	128640800
100	614*512	125747200	200	614*512	251494400
150	100*100	6120600	224	100*100	9140096
150	200*200	24240600	224	200*200	36199296
150	300*300	54360600	224	300*300	81178496
150	400*400	96480600	224	400*400	144077696
150	614*512	188620800	224	614*512	281673728

Conclusion

1. The lossless compression algorithm taking into account inter-band correlation and regression analysis allows to increase the compression ratio to (D>8)than in the use of universal archivers.

2. The proposed approach for finding the best channel groups for a given correlation value increases the efficiency of the application of the difference transformation stage.

3. The obtained experimental results show the effectiveness of the application of the stages of regression conversion and parallel processing, which allow to obtain advantages compared to analogues

References:

1. Kashkin VB, Sukhinin AI (2008) Digital processing of aerospace images. Krasnoyarsk: SFU. 278 p. (In Russian)
2. Salomon D. (2007) Data compression. The complete reference. Springer-Verlag. 1118 p.
3. Sayood K. (2006) Introduction to Data Compression. USA. Universitet Nebraska. Morgan Kaufmann Publishers is an imprint of Elsevier. San Francisco, CA 94111. 703 p.
4. William K. Pratt. (2001) Digital Image Processing: PIKS Inside, Third Edition. John Wiley & Sons, Inc, 738 p.

5. Gashnikov MV, Glumov N.I. (2016) On-board processing of hyperspectral data in Earth remote sensing systems on the basis of hierarchical compression. *Computer Optics*. vol. 40. no 4. pp. 543–551. (In Russian).
6. Gashnikov M.V. (2017) Minimization of the entropy of post-interpolation residues during image compression based on hierarchical grid interpolation), *Computer Optics*. vol. 41. no. 2. pp. 266–275. (In Russian).
7. Petrov E.P., Harina N.L., Suhih P.N. (2015) Method of fast compression of images without losses. *Machine learning and data analysis*. vol. 1. no.12. pp.1762–1770. (In Russian).
8. Petrov E.P., Harina N.L., Suhih P.N. (2016) Method of compression of multi-bit satellite images without losses. *Modern problems of remote sensing of the earth*. vol. 13. no. 2. pp. 203–210. (In Russian).
9. Petrov E.P., Harina N.L., Suhih P.N. (2016) The method of image compression in ERS systems without losses. *Bulletin of the Samara State Aerospace University named after Academician S.P. Queen*. vol. 15. no.2. pp. 183–189. (In Russian).
10. Sarinova A., Zamjatin A. and Cabral P. Lossless compression of hyperspectral images with pre-byte processing and intra-bands. Dyna (Colombia). – 2015.–Vol. 82 (190).– pp. 166-172.
11. Zamjatin A.V., Sarinova A.Zh. (2017) Algorithm for compressing hyperspectral aerospace images using mathematical processing and taking into account interband correlation. *Materials of the IV International Scientific Conference "Regional Problems of Earth Remote Sensing"*. pp.157–160. (In Russian).
12. Sergeev V.V. (2006) Analysis and Processing of Images Obtained from Earth Observations from Space. *Computer Optics*. *no*.29. pp.41–44. (In Russian).
13. Gashnikov M.V., Glumov N.I. (2014) Hierarchical GRID interpolation under hyperspectral images compression *Optical Memory and Neural Networks*. vol. 23(4). pp.246–253.
14. Gashnikov MV, Glumov NI, Kuznetsov AV, Mitekin VA, Myasnikov VV, Sergeev VV. (2016) Hyperspectral remote sensing data compression and protection. *Computer Optics*. vol. 41. no. 5. pp.689–712.
15. Gashnikov M.V., Glumov N.I. (2016) Hyperspectral image compression for transmission over communication channel. *Information Technology and Nanotechnology*. pp.334–339.
16. Hongda Shen, W. David Pan. (2017) Predictive Lossless Compression of Regions of Interest in Hyperspectral Images With No-Data Regions. *IEEE. Transactions on geoscience and remote sensing*, vol. 55. no. 1.pp.173–182.
17. Yuan L., Jianping L., Ke G. (2012) Lossless compression of hyperspectral images using hybrid context prediction. *Optics Express*. vol. 20. no.7. pp. 8199–8206
18. Changguo, Li., Ke G. (2014) Lossless Compression of Hyperspectral Images Using Three-Stage Prediction with Adaptive Search Threshold. *International Journal of Signal Processing, Image Processing and Pattern Recognition*. vol.7. no.3. pp. 305–316.
19. Mahendran M., Jayavathi S.D. (2016) Compression of Hyperspectral Images Using PCA with Lifting Transform // *International Conference on Emerging Engineering Trends and Science*. P.68–73.
20. Sepehrband F., Ghamisi P., Mohammadzadeh A., Sahebi M.P., Choupan J. (2011) Efficient Adaptive Lossless Compression of Hyperspectral Data using Enhanced DPCM. *International Journal of Computer Applications*. vol. 35. no.4. pp. 6–11.

Improvement of methods of soil survey of territories using remote sensing and 3d-modeling

Nadezhda Dobrotvorskaya[1,2], and *Alexandr* Chernov[1]

[1]Siberian state university of geosystems and technologies, Novosibirsk, Russia
[2]Siberian Federal scientific center of agrobiotechnologies, Krasnoobsk Novosibirsk region, Russia

Abstract. The effective planning and use of land resources depends to a large extent on the completeness and relevance of the digital data on the territory concerned. A modern solution for obtaining such data is the use of unmanned aircraft systems, with the presentation of the results in the form of digital elevation models and 3D-models. The article presents the practical results of the use of this technique on the example of a part of the territory with heterogeneous relief, shows its advantages with respect to the use of traditional analog data, the main directions and prospects of use for the rational use of land in land management and design of soil and environmental systems of agriculture.

Introduction. As a result of changing socio-economic conditions in Russia and the transition from a command-and-control administrative economy to a market economy, agricultural producers were faced with the need to independently plan the use of their land, based on their specific characteristics. Natural features and resource potential of land differ significantly depending on landscape conditions. The most important factor of differentiation is the terrain: differences in height above sea level, the nature of the surface, steepness, shape and exposure of slopes, the presence of forms of microrelief, etc. Depending on these factors, there are such important agronomic characteristics as heat and moisture content of soils and agrocenoses, the level of soil fertility, which in turn determine the entire technological complex of growing certain crops.

In the 60-80s of the XX century, information about the natural resources of the area was provided by the soil survey of the territory by the regional design institutes of the Giprosem system. The method of field soil survey of this period was based mainly on the use of aerial photographs as a topographical basis, as well as plans for land and forest management. Traditional soil maps can still be considered as a model of soil cover [1], however, given the gradual accumulation of variability of soil characteristics and the uncertainty of boundaries between elementary soil areas, the accuracy of such a model is limited by the subjectivity of the judgments of the soil scientist-cartographer. Topoplanes with relief lines were used mainly in areas with highly rugged terrain, as well as when conducting soil surveys of large and detailed scale for the design of reclamation (drainage and irrigation) systems.

Currently, the increased requirements for soil maps are implemented on the basis of expanding the composition of mapped features and studied dependencies, with the use of new methodological tools and techniques, including GIS technologies, with the involvement of information from other sections of soil science and related disciplines. An important direction in the development of methods for displaying the shapes of the earth's surface is 3D modeling based on the processing of aerial photography data obtained using modern unmanned aircraft systems of aircraft and helicopter types, as well as the application and subsequent automation of the procedure for processing the results of lidar survey using air laser scanners. The greatest attention in the development of this field is given to the modeling of real estate objects (especially in urban areas) [2-4], but the direction of using digital data for the design of soil and environmental systems of agriculture has not received wide coverage in the scientific and technical literature, which confirms the relevance of the study.

The object of research is selected part of the territory of LLC "Altair" Loktevsky district of the Altai territory. The site is named "Samarka". It, is located in the foothills of the Altai plain. The relief of the studied territory is characterized by a significant dissection with a predominance of surfaces with an angle of inclination of 1.5...3.0°. The area of the site is 1920 ha. Coordinates from 50.9 to 50.95° North latitude and from 81.55 to 81.63° East longitude. The map of the soil survey of 1975 was used as the initial information basis.

The method of research work included a survey of the terrain using an unmanned aerial vehicle Supercam-250, equipped with a SONY alpha 6000 24.3 MP camera, a high-precision onboard receiver JAVAD TRE-G3TAJT, then a field soil survey, the production of a relief map and a soil map.

On the basis of the obtained digital material, a digital terrain model was formed, according to which a fragment of a 3D relief model of the studied area was created. For photogrammetric processing of the results, the Agisoft Photoscan software package was used, which sequentially created a sparse and dense array of points and directly a 3D model. For clarity, the results were presented in a model area characterized by relatively large height differences (from 256 to 358 m) with an area of 52 Hectares.

Results and discussion. The accuracy of the soil survey increases significantly when using a digital terrain model (DEM). In this work, the DEM was created based on digital aerial photography from unmanned aerial vehicle with a resolution of 9 cm per pixel, using marked reference and control points. The accuracy of DEM is about 25 cm in height, that fits to the contour interval of 1 m. Construction of the DEM and DTP (digital topographic plan) was carried out using the programs Photoscan (Agisoft, Russia) and GIS-View. The DEM was presented in two versions: 1) the high-altitude part of the DTP scale of 1: 2000 in the form of pickets and horizontals through 1 meter, superimposed on the orthophotoplane. In this embodiment, shows the major forms of mesorelief that should be mapped to the DTP at an appropriate scale; 2) the elevation matrix in which each pixel corresponds to a square on the surface of the size 0,5x0,5 m This option allows the DEM to refine the surface features, including the level of the microrelief forms and surface features with dimensions of tens of meters even with slight fluctuations in elevation.

In this work, the resulting digital mapping model allowed us to detail the surface features, including the level of microrelief-the slope of the surface in degrees, the direction of surface runoff and erosion, surface shapes and elements of the size of units and tens of meters with height fluctuations up to 1.0 m (Fig. 1).

Fig.1. High-rise part of the DTP section "Samara". Isolines are drawn through 1 m. Straight lines are the boundaries of agricultural fields.

The thickening of the relief isolines reflects the location of local elevations in the studied space – hills, the soil cover of which is characterized by a low power of the humus horizon (12-15 cm) due to erosion processes, or almost complete absence of it, the output of gravelly rocks to the surface. Agricultural fields are located in spaces between the hills. There is a marked feature of the topography – the slopes are of complex shape, called accordion. This character of the slopes is particularly conducive to the development of linear erosion processes that form deep hollows and gullies in a short time.
As a result of processing the data of aerial photography of the research object, the following results were obtained:
- digital terrain model, presented as a gradient image (Fig. 2);

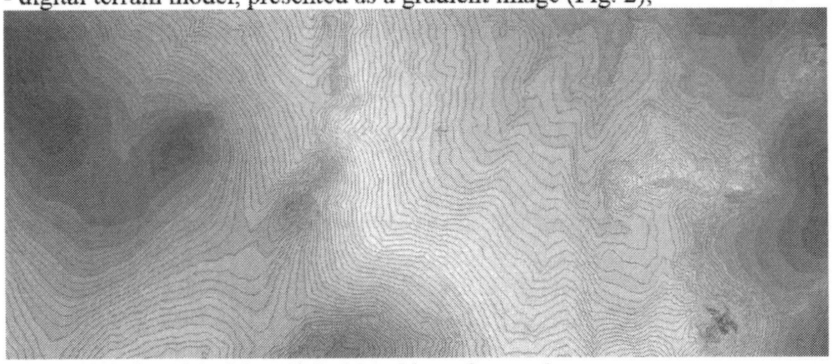

Fig..2. Digital model of the relief of the object of research

- spatial profiles that allow you to identify areas of the territory that are most susceptible to erosion processes and accordingly plan various reclamation activities (Fig.3.);

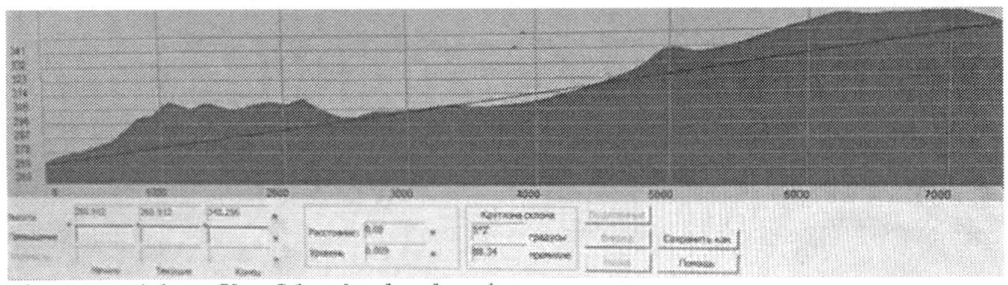

Fig. 3. Spatial profile of the simulated territory

- automatically generated statistical analysis of the simulated territory, necessary for calculating the volume of work and the economic component (Fig.4.);

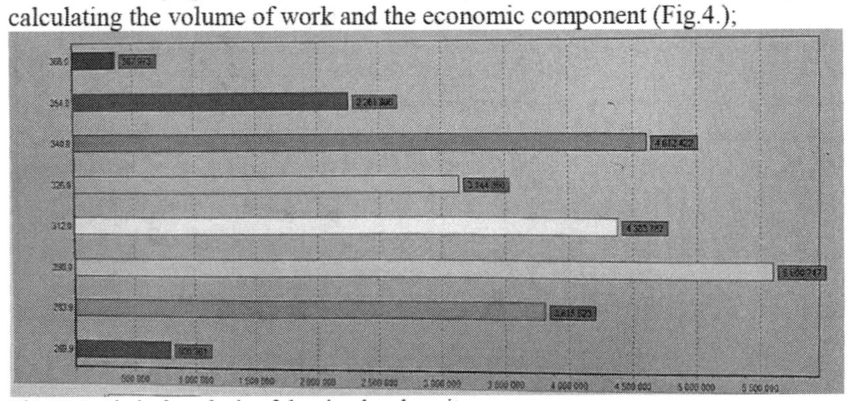

Fig. 4. Statistical analysis of the simulated territory

 https://doi.org/10.1051/e3sconf/202014902004

- 3D-model of the territory with different gradient color, clearly displaying the features and heterogeneity of the terrain (Fig. 5).

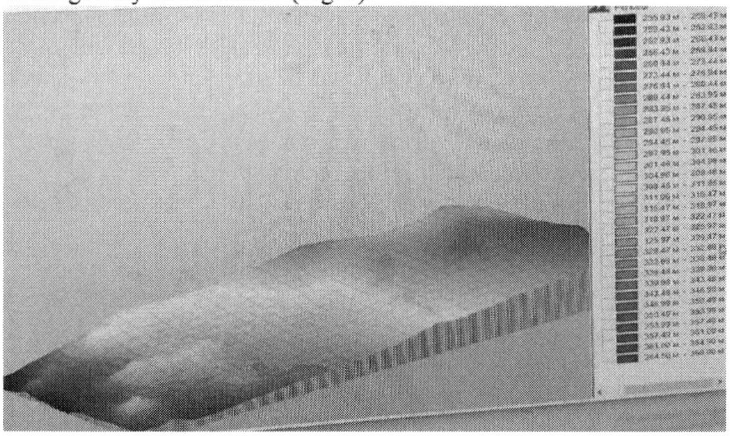

Fig. 5. 3D-model of the study area

These results allow us to identify two characteristic features. First, the heterogeneity of the slope surface and, as a result, the soil cover. It is significantly complicated by elements of micro-relief: micro-depressions and micro-hillocks, which is the cause of significant heterogeneity in the degree of soil moisture. Secondly, the model allows us to distinguish at least three high-altitude zones on the surface of the study area as a whole: high (marks in the range from 340 to 368 m above sea level), medium (from 284 to 340 m above sea level) and low (from 260 to 284 m above sea level). The boundaries between them provide greater accuracy of extrapolation of soil data obtained by the method of point soil sections. In addition, the presence of such a model significantly reduces the amount of ground soil survey associated with the identification of transition zones from one soil difference to another.

Conclusion. Modern technology, the use of digital data obtained by aerial photography as the basis for the formation of a single spatial territories, and functionality of existing software products for photogrammetric processing of results and creation of 3D-models significantly simplify the process of designing soil conservation systems of farming, organization of rational land use in land management and other fields. The use of such an integrated approach, in conjunction with the existing materials of soil and geobotanical surveys allows you to create a complete, comprehensive and up-to-date digital database of territories and automation of such work with a significant reduction in the use of human resources. In addition, the data obtained are the basis for various mathematical calculations in real time, including the volume of earthworks, land management activities, etc. The findings are demonstrated on the example of the study area, confirm the relevance and the scientific significance of the topic and are the basis for further research on this topic.

References.

1. A.N. Krenke, Correction of soil maps based on remote sensing data and digital terrain models. Digital soil cartography: theoretical and experimental studies. Sat. articles. - Moscow: Soil Institute. V. V. Dokuchaeva, Pp. 284-301 (2012).
2. A.P. Karpik, D.V. Lisitsky, K.S. Baykov, A.G. Osipov, V.N. Savinykh, Geospatial discourse of advanced and breakthrough thinking. Vestnik Sgugit, **22**, 4. Pp. 53-67 (2017).
3. A.P. Guk, M.M. Shlyakhova, Some problems of constructing realistic 3D-measurement models based on remote sensing data, Vestnik Sgugit. **4 (32),** Pp. 51-60, (2015).
4. O.A. Opritova, Study of the possible application of unmanned aircraft systems for modeling of real estate, Bulletin of SSGA, **23**, 3, Pp. 248-258, (2018).

Crown's anisotropy of a tree and scattering the signals of navigation satellites frequency band L1

Vladimir Podoprigora[1,3,*], *Anotoly* Sorokin[2,3], Daniil Makarov[2] and *Dmitry* Kharlamov[2]

[1]Institute of Physics. A.V. Kirensky, SB RAS, Akademgorodok, 50, 660036, Krasnoyarsk, Russia
[2]Federal Research Center "KSC SB RAS", Akademgorodok, 50, 660036, Krasnoyarsk, Russia
[3]Siberian Federal University, Avenue Svobodny, 79, 660041, Krasnoyarsk, Russia

Abstract. The model of interaction of signals from navigation satellites with the tree crown is considered. The contribution of the anisotropy of the dielectric constant to the cross section of radio wave scattering due to the orientational ordering of the branches is shown. The calculation of the scattering cross sections for small and large elements is presented. The tensor of orientational ordering of tree branches with division into tiers differing in the size and orientation of elements is introduced. Test measurements of navigation satellite signals passed through the crowns of spruce trees were carried out.

1 Introduction.

The study of the propagation of pulsed signals of navigation satellites (NS) in the forest is aimed at expanding the understanding of the nature of the interaction of polarized coherent electromagnetic waves (EMW) radiation in the frequency range 1-2 GHz with different-scale elements of trees and the stand as a whole. The complexity of using different models of the interaction of electromagnetic waves with the canopy is due to the dependence of the results on a large number of input parameters, including a noticeable dependence of the dielectric constant of wood on moisture and a variety of sizes and shapes of elements of the forest environment. The ratios of the wavelength of NS signals with the sizes of the diverse structural elements of the forest range from much smaller units for leaves and needles, comparable to unity for the diameters of trunks and branches, and many large units for unevenly distributed groups of trees. This leads to many interaction mechanisms and approximations partially presented in reviews [1-3]. The trunk layer has an almost ideal orientational ordering relative to the vertical. The crown layer, formed by branches of different diameters and lengths, leaves and needles, is partially ordered in space, has a biomass density and degree of ordering that varies with height. Therefore, the development of a refractive model of the stand involves taking into account the distribution of bulk density of biomass and the orientational ordering of trunks and branches of crowns.

* Corresponding author: podoprigora46@gmail.com

2 Spatial orderliness the elements of trees stand

The diversity of spatial orientation of tree branches is obvious. There is a noticeable species difference in the orientational ordering of the crowns of individual trees. The crown layer as an ensemble of volumetric structures from tree branches has several tiers with a noticeable (≈10°) difference in angles between the trunk and branches. Figure 1 shows typical crowns of birch and larch. The description of orientational ordering is presented in sufficient detail in the physics of molecular systems.

Fig. 1. Typical crowns of birch and larch.

In the theory of liquid crystals and polymers, the orientational ordering of molecules is described in the general case by a tensor [4-5]. The elements of the orientational ordering tensor S_{ij} of the tree have the form:

$$S_{ik} = \frac{1}{N-1}\sum^{N-1}\frac{1}{2}\left(2e_{\alpha}^{i}e_{\beta}^{k} - \delta_{ik}\right), \qquad (1)$$

where N is the number of branches, e^i_α and e^k_β - components of the unit vector along the directions of the α-th and β-th branches, δ_{ik} is the Kronecker symbol.

Three eigenvalues of the tensor S are orientational ordering parameters, and the maximum eigenvalue corresponds to the order parameter S, which is standardly defined in the theory of liquid crystals and polymers as the average value of the second Legendre polynomial in angles θ between the direction vectors of the selected direction (vertical) of the trunks and branches.

In the crowns of different types of trees, the orientational orderings of branches by tiers differ markedly. Table 1 shows the different types of tiers and the corresponding types of tensors of orientational ordering of branches. Tiers are counted from the top of the tree to the soil.

The data on the type of tensor allow us to take into account the contribution of crowns to the vertical dependence of the real part of the anisotropy of the effective dielectric constant of the stand, which has values in the range $0 < \Delta\varepsilon' < 0 < \Delta\varepsilon' < (\varepsilon_l' - \varepsilon_c')$. The indices L and C denote the longitudinal (Long, along the trunk) and transverse (Cross) dielectric constant of wood.

Model calculations of the real part of the effective dielectric constant of crowns are carried out taking into account the volume of wood in crowns and the eigenvalues of the tensor S_{ij}. orientational order parameter. There are experimental data on the dielectric constant of wood with a humidity of 40% at a temperature of 20°C at frequencies of 0.915 and 2.375 GHz [6]. Field measurements of the orientation of the main branches were carried out according to photographs during the year of trees of various species from 20 to 60 years

 https://doi.org/10.1051/e3sconf/202014902005

old from a distance of 30-70 meters. Data were obtained on the orientational structures of birch and larch without foliage and needles, respectively. The orientation parameter of the order of the main branches of the crowns S decreases from the top to the lower border in the following intervals: birch 0.8 - 0, larch from 0.5 to - 0.4. The order parameters of the branches of all types of trees refers to the main branches. The interval S with positive values only has a birch.

Table 1. Types of crown levels and type of tensor S_{ij}.

Type of tier, angle	$\langle s \rangle$	Type of tier, angle	$\langle s \rangle$
I , $\theta = 30^0$ 	$\frac{1}{8}\begin{pmatrix}1 & & \\ & 2 & \\ & & 5\end{pmatrix}$	II , $\theta = 60^0$ 	$\frac{1}{8}\begin{pmatrix}2 & & \\ & 2 & \\ & & -1\end{pmatrix}$
I , $\theta = 90^0$ 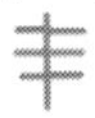	$\frac{1}{8}\begin{pmatrix}2 & & \\ & 2 & \\ & & -4\end{pmatrix}$	IV, $\theta = 120^0$ 	$\frac{1}{8}\begin{pmatrix}2 & & \\ & 2 & \\ & & -1\end{pmatrix}$

Estimates of the numerical values of Δε' for the layers of the three upper tiers of the crowns of the test stands of birch and larch are made. Changes in the tiers of the orientation parameter of the order of branches and density of wood are taken into account.

Table 2 presents the calculated values of the dielectric constant anisotropy Δε of three tiers of birch and larch crowns.

Table 2. Anisotropy of crown stages taking into account the order parameter **S**.

Tiers of crowns	Birch			Larch		
	$\Delta\varepsilon$, S = 1	S	$\Delta\varepsilon$·S	$\Delta\varepsilon$, S = 1	S	$\Delta\varepsilon$·S
1 tier	0,002	0,6	0,001	0, 004	0,4	0,002
2 tiers	0,025	0,3	0,008	0,026	0,2	0,005
3 tiers	0,24	0,1	0,02	0,23	0,1	0,02

3 Attenuation of NS signals in the tree stand

Typical structural characteristics are inherent in any tree of a certain type and age: the shape and size of the crown, the height and diameter of the trunk, the location and orientation of large and small branches, leaves and conifers. In this paper, the tree model is described by a set of canonical elements distributed over different layers (tiers). Tree trunks and branches are modeled by cylinders of finite length, leaves and needles are represented by dielectric ellipsoids. For the radio frequencies of navigation satellites (1.2-1.6 GHz), the elements of the tree are quite transparent, the transmitted and scattered radiation reacts to the anisotropy of the wood and the orientational ordering of trunks and branches [1].

The derivation of expressions for scattering of radio waves by dielectric ellipsoids is based on the generalized Rayleigh-Hans-Debye approximation (RHD), which requires the fulfillment of the constraint:

$$k_0 d|\sqrt{\varepsilon} - 1| << 1, \qquad (2)$$

where $k_0 - \frac{2\pi}{\lambda}$ - free space wave vector, d – smallest axis of an ellipsoid, ε - the dielectric constant. Expression (2) implies that the diffuser partially transmits the probe wave.

Since the sizes of large tree elements (trunk, branches) in the considered L-frequency range are comparable with the radio wavelength (λ ≈), scattering cross sections at arbitrary incidence angles were calculated using the diffraction model for infinite and finite cylinders [2,3]. In this model, the only restriction on cylinder size is:

$$h >> r, k_0 h >> 1, \quad (3)$$

where h and r – cylinder height and radius, respectively.

It is known [5] that the scattering of electromagnetic waves can be interpreted in terms of low-amplitude fluctuations in the orientations of scattering objects. In this case, the differential scattering cross section σ per unit of scattering volume in a unit of solid angle per unit of change in the angular frequency ω is written as:

$$\frac{d^2\sigma}{d\Omega d\omega} = \pi^2\lambda^{-4}\int\int_{-\infty}^{+\infty}\langle\delta\varepsilon^2\rangle\, exp[i(\overline{qr} - \omega t)]\, drdt \quad (4)$$

where $\lambda = \frac{2\pi c}{\omega_0}$ – wave length in vacuum, $\overline{q}$- wave vector, $\langle\delta\varepsilon^2\rangle = \langle\delta\varepsilon(0,0)\delta\varepsilon(r,t)\rangle$ - mean square permittivity at a given point at a given time *t*.

For the wave vector q and frequency ω, the differential scattering cross section per unit volume has the form [5]:

$$\frac{d\sigma}{d\Omega} = \frac{\pi^2}{\lambda^4}\langle\delta\varepsilon^2\rangle_q = \frac{4\pi^2}{q\lambda^4}(\Delta\,\varepsilon)^2\langle S_{ij}^2\rangle_q \quad (5)$$

where $\delta\varepsilon = \varepsilon_{ij} - \overline{\varepsilon_i}$, $\overline{\varepsilon}$ - average dielectric constant, $\Delta\varepsilon = \varepsilon_{||} - \varepsilon_{\perp}$ - anisotropy ε between the directions along and across the barrel in the case when the diffusers are parallel to each other, S_{ij} - components of the orientational ordering tensor. Formula (5) is valid if the imaginary part ε is much smaller than the real part.

Tensors $\langle S_{ij}\rangle$ and $\langle S_{ij}^2\rangle$ can be calculated both for small twigs and needles oriented at certain angles to large branches, and for the main crown-bearing branches and the whole tree, divided into tiers. The tier contains areas of branches with the same angle of inclination θ from the vertical. This angle from tier to tier varies from 20° to 120°. The off-diagonal elements of the tensor are equal to zero due to the axial symmetry of the crown. The off-diagonal components of the tensors $\langle S_{ij}^2\rangle$ do not generally vanish and can appear in the elements of the scattering tensor during polarization measurements.

Experimental data on the dependence of the elements of the tensor of the complex dielectric constant ε = ε' + *i*ε" of wood on humidity and temperature for radio frequencies in the range of $20\text{-}10^{11}$ Hz are presented in the monograph [6]. The difference in the dielectric properties of wood is observed for the orientations of the electric field vector along the fibers (trunk), in the radial and tangential directions. In practical applications, tensor elements corresponding to the longitudinal and transverse directions of the wood are used. Crowns have a vertical length less than the trunks. The share of crown biomass is 0.1-0.2 of the volume of the tree. The diagonal elements of the orientational ordering tensor S_{ij} of branches vary from upper tiers to lower. This circumstance, taking into account the area of interpenetration of the layer of trunks and crowns, makes it possible to make model estimates of the vertical dependence of the effective dielectric constant of the stand as a whole.

4 Radioscopy of crowns.

A coherent pulse signal on the surface of the Earth is practically a plane wave. The NS signal receiver provides registration of the angular coordinates of the wave vector of the signal. The attenuation of the signal in the stand is due to three factors: absorption, scattering on the elements of the crowns and trunks, the difference in phase delays of the vertical and horizontal components of the electric field.

The direct NS signal recorded by the receiver passing outside the crown is stably reproduced in the absence of multipath. The passage of the probe signal through the crown leads to changes in the amplitude of the signal at the receiver, recording as direct and scattered and reflected signals by the trunks and elements of the crowns. Spatial heterogeneity of the effective dielectric permittivity of the tree stand generates multipath signals at the receiving antenna with a continuous distribution of phase shifts.

In the conditions of neighboring crowns radioscopy, it is possible to select a test option from two trees with close crowns. The radio transmission scheme of two fir trees with developed crowns is shown in Figure 2. The trees are located at a distance between the trunks of 3.4 m. The height of the trees is 24 and 25 m, the diameter of the trunks is 0.32 m. The crowns are conical in shape, with a height of 22 meters. The lower part of the crowns of both trees partially closes.

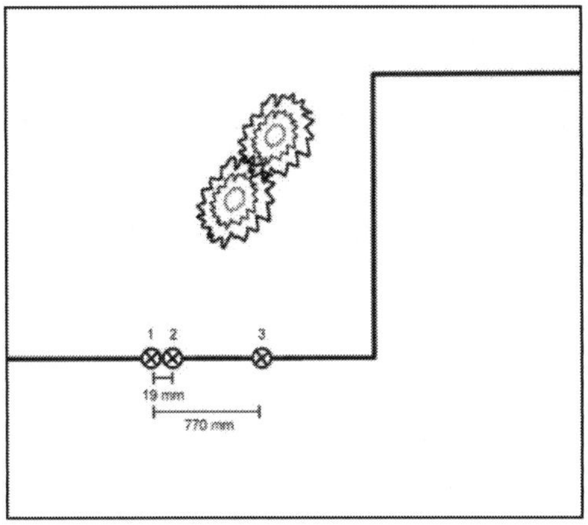

Fig. 2. Scheme of radioscopy of crowns of two fir trees.

The pulsed radiation of NS signals with right circular polarization during passage through the crowns of fir trees is attenuated as a result of scattering and absorption and partially depolarized in the volume occupied by the trees. The depolarization effect is determined by the anisotropy of the effective dielectric constant of the crowns of the trunks. The interference effects of coherent radiation in the forest stand form an EMW of a complex spatio-temporal structure.

Transmitted NS signals were recorded by MRK-32R signal receivers, manufacturer of NPO Radio Communications, and specialized receiver-recorder NSRP-04, manufacturer of Phoenix Engineering Bureau LLC, Krasnoyarsk. A set of antennas was used to receive signals with linear and right circular polarization.

Figure 3 shows the spatio-temporal dependences of the amplitude of NS signals located relative to the antennas in the intervals of azimuth angles α (-30 ° - 120 °) and elevation angles φ (-20° - 60°). Antennas 1 and 3 received signals of circular polarization right, 2 - linear, oriented vertically. The axes are the amplitudes of the received signal and the number of registration frames. In the array of data recorded by the receiver for each frame, the date, time, and angular coordinates of the NS are indicated. The frame rate is 1 Hz.

The results of scanning crowns in different angles from different angles demonstrate spatial heterogeneities of the effective dielectric constant of the volume occupied by two fir. Further experimental studies and variants of NS signal interaction models suggest polarization measurements, the use of different projections of radioscopy in order to restore the spatial structure and gradients of the effective dielectric constant of the stand.

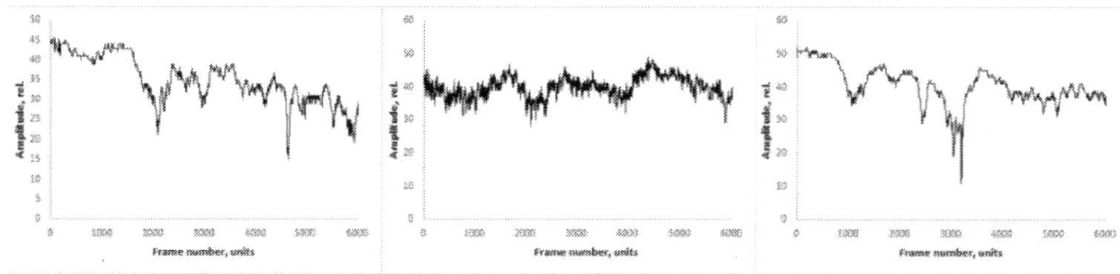

a) b) c)

Fig. 3. Dependences of the amplitudes of NS GPS 15 signals attenuated in the crowns of fir trees from antennas: a) – 1, b) – 2, c) – 3.

5 Conclusion

A model of the interaction of signals from navigation satellites with a tree crown is considered. The contribution of the dielectric constant anisotropy to the scattering cross section of radio waves due to the orientational ordering of the branches is shown. The calculation of scattering cross sections for small and large elements is presented. The orientational ordering tensor of tree branches is introduced with the division into tiers differing in the size and orientation of the elements. Implemented the option of using scanning of tree crowns in different directions by signals of navigation satellites. Informative differences of the signal amplitude at the location of the receiving antenna from the paths of the emitting antenna. Variations in the amplitude of the recorded signals are associated with the spatial distribution of the effective dielectric constant of the tree, depending on the volume fractions and humidity of the wood substance of the trunk and crown. The use of radioscopy in a wide range of azimuth and elevation angles of navigation satellites has the prospect of developing radio tomography of the volumetric dielectric structure of a tree crown.

References

1. V.I. Popov, Propagation of radio waves in forests - M.: Hot line - Telecom,.- 392 p. (2015)
2. A.Chukhlantsev, A. Shutko, S. Golovachev. Radio engineering and electronics, 48, 11, 1285 (2003).
3. Ulaby F. T., Tavakoli A., Senior T.B.A. // //IEEE Trans 1987. V. GRS-25. P.714-719
4. Blinov L.M., Liquid crystals: structure and properties. M.: Book House <LIBROCOM, 2013. - 480 p .; col. incl.
5. Computer simulation methods for the study of polymers and liquid crystals (responsible editor V.A. Ivanov) Moskva, Izdatelstvo "Mir",. «Librocom». - 696 p. (2009)
6. Torgovnikov G.I. Dielectric Properties of Wood and Wood Based Materials. – Berlin, Springer – Verlag. – 1993. – 194 p.
7. S.N. Novik, A.S. Zavyalov, E.D. Telpukhovsky, Electrophysical and dissipative characteristics of the properties of the elements of a coniferous stand // Izvestiya Vuzov. Physics. **49**, 7, P.57-67. (2006)

Assessment volume of a forest biomass by the attenuation of the navigation signals the frequencies L1

Anatoly Sorokin[1,2], *Anatoly Ostylovsky*[2], *Aleksandr Borisov*[3], *Viktor Ivanov*[3], *Daniil Makarov*[1], *Dmitry Kharlamov*[1]

[1]Federal Research Center KSC SB RAS, Akademgorodok, 50, Krasnoyarsk, 660036, Russian Federation
[2]Siberian Federal University, 79, Svobodny, Krasnoyarsk, 660041, Russian Federation
[3]Institute of Forest, KSC SB RAS, Akademgorodok, 50, Krasnoyarsk, 660036, Russian Federation

Abstract. The method for assessment the volume of a forest stand based on measurements of the attenuation of navigation satellites signals in the L1 range and morphological features of forest species is presented. The dielectric model of a mixed dielectric with the natural structural features of the species of the forest stand was used. The option of restoring the biomass volume of the forest stand according to the radioscopy data on the mixed forest test site is considered. The practical applications of the method of continuous local monitoring of the forest stand condition in forest science are discussed.

1 Introduction

The physics of the interaction of radio waves with the forest is an actual area of research in fundamental and applied sciences. Forests are randomly heterogeneous environments with tree elements of various shapes, sizes and spatial orientations [1]. The electrophysical properties of the stand as a multicomponent medium are determined by the molecular structure of wood, including water, and the macroscopic distribution of tree elements in the volume of the stand.

It is possible to continuously monitor the state of the stand by radioscopy using signals from navigation satellites [2]. The electrophysical characteristics of the stand are determined by the amount of stand biomass by fractions, the moisture content of wood, needles, leaves and the spatial distribution of trees.

The sounding of forests with signals from various satellites of the GPS and GLONASS navigation systems with a frequency of 1 Hz provides scanning of the forest stand on an area of up to 1 ha. Coordinate binding of scanning flows of pulse signals allows determining the localization of changes in biomass and its moisture [3].

2 Method for determining the bulk density of stand biomass

The signals of each navigation satellite cover half the surface of the globe. The location of the antenna in the tree stand on the soil surface with a vertically oriented axis of the radiation pattern makes it possible to record the probing flux of an individual satellite in the analytical volume V in the form of a cone with an angle of 160 ° and a vertex on the antenna. In this volume, we distinguish conical channels with a solid angle corresponding to the first Fresnel zone. In this volume, the antenna registers signals scattered by the forest stand with a path difference of not more than $\lambda/4 \approx 5$ cm. The contributions of signals with a greater difference in stroke will be noticeably smaller and this volume mainly determines the value of the recorded signal after passing through the stand. The azimuth and elevation angle of the satellite, in conjunction with the coordinates of the antenna's location, allow us to calculate the coefficients of specific attenuation in the stand with coordinate reference [3].

Tree stand as a statistically heterogeneous mixed dielectric medium with local conductivity for radio waves of the L1 range is a scattering and absorbing medium. The total signal power loss during propagation in the forest is determined by absorption and multiple scattering on wood, needles, leaves [1].

In the volume V_F of the conical channel with the solid angle of the first Fresnel zone in the layers of small thickness δ, perpendicular to the wave vector of the signal, with the volume fraction of wood μ and attenuation coefficient ν, the signal power loss will be determined by the product νμδ.

The passage of the signal from the upper edge of the stand to the antenna through the number of layers H / δsinφ gives the attenuation coefficient of the signal $(1 - \nu\mu\delta)^{H/\delta\sin\varphi}$. The passage to the limit as $\delta \to 0$ leads to the expression for attenuating the signal power P_0 at the input to the stand to the value P at the receiving antenna:

$$P = P_0 \exp\left(\frac{\nu\mu H}{\sin\phi}\right) \quad (1)$$

The obtained exponential law is similar to the known laws of absorption and scattering of electromagnetic waves of the optical range in various media [4]. The volume V_F is determined from the angular coordinates of the navigation satellite and the average height of the stand. The average value of the fraction of the volume of biomass in this volume is determined by the expression μV_F, where:

$$\mu = \frac{\ln\frac{P_0}{P}}{\nu H}\sin\phi \quad (2)$$

3 Radioscopy of the forest stand.

The key point to identify the possibilities of the method of determining the volume of the stand is the selection of test objects. The characteristics of the transmitted signals of navigation satellites in the forest stand vary depending on the species composition, age of the trees, the diameter of the tables, the size and structure of the crowns, the structure of the distribution of trees in the study area [4]. Carrying out measurements, data analysis and identifying the relationship between changes in the characteristics of transmitted radio waves through the forest canopy with specific biometric indicators of the stand requires the maximum reduction in the number of factors infuencing the attenuation of signal power. Such objects can be stands that are clean in composition with uniform spatial distribution of trees, without glades and clearly defined grouping. Desirable absence of undergrowth, undergrowth and shrub layer. Important is the pronounced layers of crowns and trunks of the forest stand.

However, mixed coniferous-deciduous forests are interesting in that they have noticeable changes in the electrophysical characteristics of crowns associated with the appearance and decay of foliage against the background of persisting needles. Therefore, as the first test object, a mixed pine-birch plantation plot located on the territory of urban forests near SFU was selected. A satellite image and a fragment of the stand are presented in Fig. 1.

Planting is represented by small areas of pine and birch forests. Pine forest stand (small grass pine stand) with an area of about 1.0 ha of class II bonitet has the following taxation characteristics: diameter at a height of 1.3 m 26.0 cm, height 23.5 m, fullness 0.9; stem wood stock - 290 m^3. Birch forest stand (motley grass birch stand) with an area of 0.4 hectares of class III bonitet has the following taxation characteristics: diameter at a height of 1.3 m 25.0 cm, height 20.5 m, fullness 0.7, stem wood stock 170 m^3.

Fig. 1. A fragment of the stand of pine and birch plantations.

Teenage, undergrowth and shrub layer are absent. Planting is represented by small areas of pine and birch forests. Pine stands with an area of about 1.0 ha of class II bonitet has the following taxation characteristics: diameter at a height of 1.3 m 26.0 cm, height 23.5 m, fullness 0.9, stem wood stock 290 m^3. Birch forest stand with an area of 0.4 ha of class III bonitet has the following taxation characteristics: diameter at a height of 1.3 m 25.0 cm, height - 20.5 m, fullness - 0.7, stem wood stock - 170 m^3.

The structure of the layers of crowns and tree trunks is physically significantly different in the density of distribution of wood substance in the volume of the stand. The effective dielectric constant will also be different. The attenuation of the signals of navigation satellites in the forest stand is determined by two main taxation indicators - the sum of the cross sections of trunks in the lower part of the stand at a height of 1.3 m and the biomass of tree crowns. These indicators can be obtained from the results of taxation using patterns of distribution of biomass in the stand. The calculation of the sum of the cross-sections of the trunks is quite simple to perform using the distribution of trunks by their diameter, and to estimate the biomass of the crowns it is necessary to use a rather complicated technique. The study of biomass is associated with the study of the interconnections of its various fractions in model trees. There are a large number of publications on the relationship of the mass of fractions of individual trees with a diameter of 1.3 m, height, cross-section of the trunk, as well as age [5-6].

https://doi.org/10.1051/e3sconf/202014902006

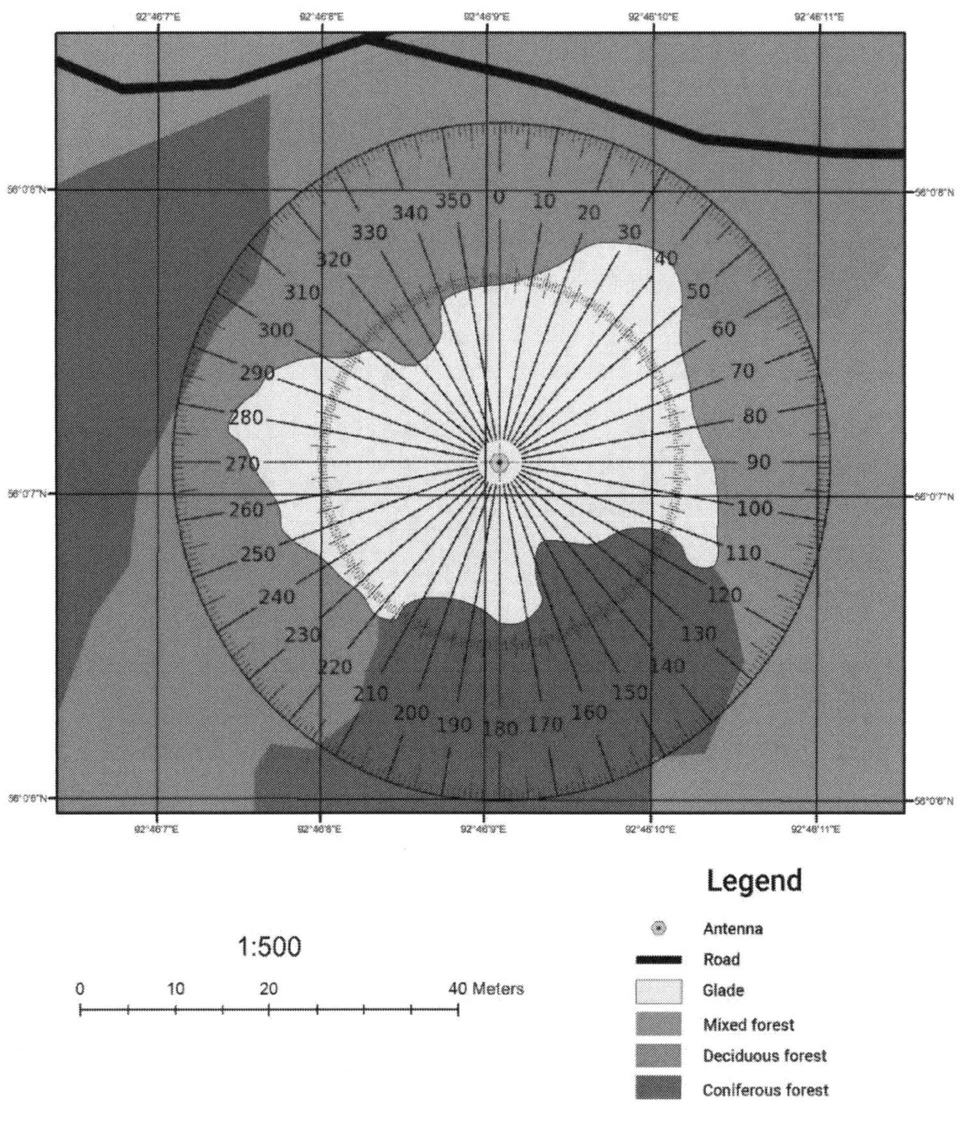

Fig. 2. Placement of mixed forest tree species on the test site.

Signals of the L1 range of GPS and GLONASS systems were recorded with MRK-32 equipment. The height of the antenna above the soil level is 0.25 m. The axis of the antenna pattern is oriented vertically. Antenna location coordinates 56°00'07.6"N 92°46'09.2"E. The antenna was in the middle of the edge with a shift of 3 m to the southern border of the site.

The duration of the registration session is 3 hours. The sampling frequency was 1 Hz; during the indicated time interval, the forest massif scanned 13 satellites from each constellation. The location of the antenna provided the registration of GLONASS, GPS signals that passed through the forest and partially free space in the range of azimuth angles 0-360° and elevation angles 10-90°. The maximum length of the signal path in the forest is about 100 m. Graphs of changes in signal amplitude over time (frame scale, 1 s resolution) are shown in the figure 3.

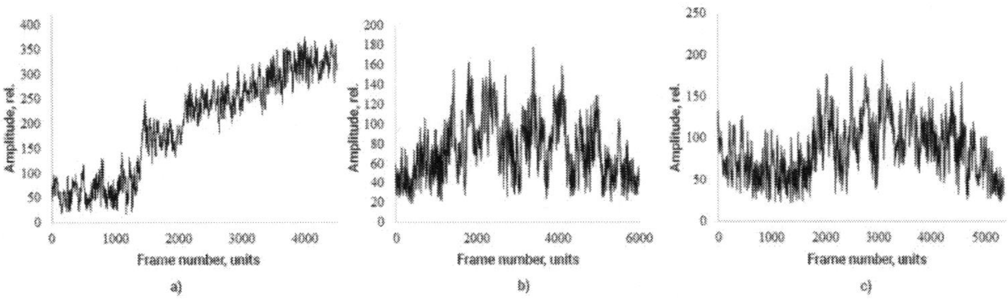

Fig. 3. Graphs of changes in signal amplitude over time: a) GLONASS 18 satellite, forest (frame numbers 1 - 2000) and free space (frame numbers 2000 - 4000), elevation angle ranges: 19°-55° and azimuths: 230°-260°; b) GPS 3 satellite, forest range elevation angles: 16°-34°, azimuths: 100°-37°; c) GPS 24 satellite, forest, elevation ranges: 21° - 31°, and azimuths: 260°-320°.

4 Results and discussion

The uniquely determined position of the analytical volume VF of the conical channel allows revealing spatial dielectric inhomogeneities in the stand associated with variations in biomass and its moisture content. Temporal dependences of the amplitude of attenuated signals in the forest stand are used to determine the coefficients of specific attenuation assigned to the areas of different types of forests taking into account the known taxation characteristics. The sounding signal for satellite trajectories with elevation angles of 20-40° passes successively layers of crowns and trunks. Increasing the elevation angle of the satellite allows you to select signals attenuated only by the crowns, changing the azimuthal angle provides scanning of the stand area to 3 ha in the range of angles 0-360°. The associated values of ν and μ, if it is possible to calibrate using homogeneous stands, make it possible to restore ν at a known value of μ determined by taxation characteristics. In long-term monitoring of the stand, it is necessary to take into account changes in the dielectric constant, associated with seasonal cycles of moisture in the wood and needles, and the state of foliage. [7]

5 Conclusions

The fundamental possibility of the method of recovering fractions of the volume of wood, foliage or needles requires the development of a dielectric model based on data on the dielectric constant of wood, morphological and structural characteristics of the stand, and calibration of the probe signal. In reality, the creation of specialized receivers with software that allows you to process arrays of recorded signals to a user level within 3-4 hours. Placing the receiving antenna at different heights above the soil provides data on the crown layer. Practical applications of the method of continuous local monitoring of the stand state by signals of navigation satellites can be useful in forestry and forest pathology.

References

1. V.I. Popov, Propagation of radio waves in forests - M.: Hot line - Telecom,.- 392 p. (2015)

2. V. L. Mironov, M. I. Mihaylov, A. V. Sorokin, K. V. Muzalevskiy, S. V. Fomin. Method of measure of microwave radiation at the forest canopy using GLONASS- and GPS-signals // Vestnik SibSAU. T. 51. – № 5. – pp. 123-126, (2013)
3. Pat. 2682718 Russian Federation, IPC G01S 19/03 (2010.01) Method for determining the linear attenuation coefficients of navigation spacecraft signals in a forest with coordinate reference / Makarov D.S., Savin I.V., Sorokin A.V., Fomin S.V. ., Kharlamov D.V. No. 2018119448; declared 05/25/2018; publ. 03/21/2019, bull. No. 9.8 p.
4. A.V. Sorokin, V.G. Podoprigora, S.V. Fomin, I.V. Savin, Evolution of the signal characteristics of navigation satellites in the forest. Izvestiya vuzov. Physics.. 60. 12/2. pp. 19-22, (2017)
5. Babich N.A., Travnikova G.I. The structure of the phytomass of pine forests of artificial origin of the boron ecological series // Forest Journal 1990, No. 3. S. 10-16.
6. Stakanov V.D. The distribution of organic matter in various parts of Scots pine // Forestry, 1990. No. 4. S.25-33.
7. Torgovnikov G.I. Dielectric Properties of Wood and Wood Based Materials. – Berlin, Springer – Verlag. – 1993. – 194 p.

Geological interpretation of the cosmic materials from the South of Western Siberia in connection with the petroleum potential of the sedimentary cover

Andrew Belonosov[1,2], *Anton* Kudryavtsev[2], *Sergey* Sheshukov[2], and *Dmitry* Borisov[3]
[1]Tyumen Industrial University, 625000, Tyumen region, Tyumen, Russia
[2]West Siberian branch of the Trofimuk Institute of Petroleum Geology and Geophysics of Siberian Branch Russian Academy of Sciences (WSB INGG SB RAS), laboratory « Geology of oil and gas», 625026, Tyumen region, Tyumen, Russia
[3]V.I.Shpilman research and analytical centre for the rational use of the subsoil, 625026, Tyumen region, Tyumen, Russia

Abstract. In the South of Western Siberia oil-perspectivity Jurassic deposits are characterized by multi - and small-scale. The interpretation of earth remote sensing materials in the visible, near and far infrared ranges allowed to evaluate the oil potential of numerous domes and depressions on the basis of combining geodynamically stressed zones and calculating the physical characteristics of the earth's surface (albedo, radiation coefficient, thermal inertia, convective heat flow, daily evaporation of moisture, DEM, weather conditions, etc.), including the procedure of reference classification, where the standards are the nearest oil and gas condensate fields. The forecast boundary of "oil and gas condensate" lands of the South of Western Siberia is displaced to the latitude of u.v. Lebyazhye of the Eastern part of the Kurgan region.

Introduction

The geography of oil and gas searches is expanding every year. Despite the fact that the main vector of prospecting geological and geophysical research is directed to the northern and eastern outskirts of Western Siberia, the western and southern parts still belong to reserve resource territories [1]. With the accumulation of new materials of geological and geophysical work, a regional reassessment of the oil potential of poorly studied large structural and tectonic elements is periodically carried out: anteclise, syneclise, anticlines, depressions, domes and hollows [2-3].

In the south of Western Siberia, including Tyumen (without autonomous okrugs - Khanty-Mansi Autonomous Okrug and Yamalo-Nenets Autonomous Okrug), Kurgan and, partially, Omsk and Sverdlovsk regions, according to the results of regional and fragmented areal seismic surveys, taking into account about 300 exploration wells already drilled, a reduction in sedimentary cover southward. In the border zone with the Republic of Kazakhstan, its thickness is on average 50-100 meters.

Currently, the border of oil-prospective lands in the south of Western Siberia is conditionally drawn along the administrative border of the Uvat and Vagai districts of the Tyumen region.

Relevance

On the other hand, the pre-Jurassic base (conventionally Triassic-Paleozoic basement) of the West Siberian Plate (WSP) has a stepped structure and plunges to the north, and oil-prospective Jurassic sedimentary cover deposits are characterized by multi-dome and small dome [3-5]. Given the small size of the structures, and, accordingly, the small reserves of hydrocarbon raw materials, the question arises of how to classify them into productive and "empty" ones, since it is not possible to verify each small anticline structure with a deep (about 3 km) exploratory well. Therefore, leading oil companies are reluctant to consider the oil prospects of the southern regions of Western Siberia, and medium and small companies, as a rule, do not have enough financial support for the full cycle of exploration. From here, the following problems become relevant:

1) to offer and interest leading oil and gas companies with large prospective geological objects;

2) apply the methodology for remote forecasting of "oil" lands, after conducting oil prospective zoning and filtering of "empty" and productive structures.

Problem state

The materials of regional seismic studies taking into account borehole materials show that several large depressions are identified in the sedimentary cover in the south of the WSP, which correspond to submeridional graben-like extended structures of the pre-Jurassic basement [6-8].

On the one hand, graben-like depressions are characterized by a complete set of Mesozoic sedimentary deposits from the Triassic to Cretaceous. On the other hand, graben-like basement structures are limited by deep faults, which are translators of hydrocarbon fluids to the earth's surface [9]. Thirdly, debris demolition material is accumulated in the depressions, which is usually a good collector. Fourth, the presence of the Bazhenov horizon is a high-quality cover, which sharply increases the likelihood of the formation of oil deposits [2].

Due to the insufficiency of geological and geophysical materials, various packages of Earth remote sensing (ERS) materials find their place in the detailed regional oil exploration work.

To decrypt these packets, not only «classical» pattern recognition can be used, but also advanced software products using neural networks.

Satellite technologies make it possible to assess, to a first approximation, the oil potential of depressions for setting up further onshore exploration work [3-5, 10-12].

Problem solving methods

To solve the above problems, space materials were used in the visible, near and far infrared ranges [10-12]. These are fragments of space images at a scale of 1: 500 000 (a component of the Geolart-1000 remote base, a composite based on satellite imagery materials from KATE-200 cameras - spectral zones 510-850 nm), fragments of a global mosaic taken by an ETM + scanner of «Landsat-7» satellite in the spectral range of 0.53-0.9 and 1.55-1.75 microns, as well as a digital elevation model (DEM) - SRTM DEM based on data from the SIR-C and X-SAR sensors of the Shuttle spacecraft.

The methodology of geological analysis and decoding of space materials consisted of three stages [3-4].

The first stage – «Processing and decoding of space images» - involved topographic and geodetic linking of space images, geomorphological studies and lineament analysis. Further, when generalization of the lineament pattern, geodynamically stressed zones (GDNZ) were identified. They represent a «framework of mobile tensile structures» and characterize regional deep fracturing, as well as fluid dynamic processes in the sedimentary cover.

The second stage – «Integration of geological and geophysical materials with GDNZ systems» - consisted in a comparative analysis of the GDNZ framework with geophysical and geochemical fields, with structural maps, with geological sections, data from exploration wells, etc. etc.

The third stage - «Geological interpretation of space materials» - consisted of examining the processes of oil and gas accumulation in the traps of hydrocarbons in the framework of the GDNZ.

An interesting fact is that hydrocarbon traps with proven oil and gas potential are within a certain radius from the intersection of the GDNZ (nodes GDNZ). On the day surface, the nodes of GDNZ are characterized by anomalous positive or negative effects of geophysical, geochemical, thermal, radioactive, microbiological and other fields.

Large depressions of the sedimentary cover that inherit the general structure of a series of grabens of the foundation, as a rule, coincide in plan with the GDNZ highlighted from space materials.

An analysis of the regional distribution of hydrocarbon deposits on remote sensing materials showed that most oil and gas fields are located near linear thermal zones. These zones, as a rule, display active fluid-conducting transit faults, leaving the foundation and cutting the entire sedimentary cover to the earth's surface. On maps of a larger scale, oil and gas deposits are confined to local thermal anomalies, or to their «rings». This is confirmed by the verification of the ground geothermal and geochemical surveys. The explanation of this fact is the vertical process of heat and mass transfer by the fluid-conducting systems of cracks and faults, delineating hydrocarbon traps [9].

Thus, thermal space survey (TSS) can serve as the first step in determining the oil and gas potential of structural and tectonic elements of the upper part of the sedimentary cover in poorly studied territories.

But it should be noted that not all thermal anomalies are a direct criterion for oil and gas saturation of sedimentary deposits.

In order to systematize space thermal anomalies (which of them are associated with hydrocarbon deposits, and which are not), a methodology was developed at the Center for Environmental Safety of the Russian Academy of Sciences of the Russian Federation (St. Petersburg) based on the use of multiple TSS and a special algorithm for processing thermal images of the earth's surface (ES). The methodology included the calculation of the physical characteristics of the earth's surface, such as the total solar radiation flux density, albedo, emissivity, thermal inertia (TI) ES, convective heat flux (CHF) from the subsoil, daily moisture evaporation (DM) ES, soil moisture, digital relief model (DEM) ES, soil structure, meteorological conditions, the concentration of optically active gases (CO2, SO2) in the atmosphere and time [10, 12].

The calculation algorithm for CHF, TI and DM was based on a mathematical model of the daily course of the Earth's surface temperature (EST), taking into account weather data (temperature, pressure, air humidity, wind speed, cloud cover) and the main factors under the influence of which the EST is formed.

For zoning of the territory of the South of Western Siberia, digital mosaics were prepared using the Geomatics and ERDAS software tools based on the quantitative characteristics of the earth's surface.

The interpretation of space materials with the aim of forecasting oil potential lands was carried out through the classification procedure of the data obtained above. For the reference classification, the maximum likelihood method was applied [10].

As a training sample, oil, oil condensate and gas condensate fields used in the GIS "Natural Resources of Russia" were used. For analysis, several standards were selected, representing a set of features of the ES in already known oil and gas fields located in the vicinity of the studied territory.

The discussion of the results

As a result, from the selected standards (deposits) in the South of Western Siberia right up to the border with the Republic of Kazakhstan, spatially recorded places (areas) where indications from the Kalchinsky, Polunyakhsky, Taytymsky and Tevrizsky oil, oil and gas and gas condensate fields appeared.

The spatial distribution of these places has a clear pattern. Starting from the latitude of Tobolsk and to the state border with the Republic of Kazakhstan, the number and area of these promising places is gradually reduced, and in the border areas it is completely disappearing. This trend is logical, since the thickness of the sedimentary cover, for which the standards (deposits) were chosen, is practically reduced towards the border with the Republic of Kazakhstan.

When applying a structural map along the roof of the Bazhenov horizon with oil potential areas identified using the reference classification according to remote sensing data, approximately half of the small structures were classified as unpromising. Thus, the number of promising anticlinal structures underwent the first stage of filtration and their number sharply decreased, almost 2 times.

Conclusion

According to the remote forecast of hydrocarbon deposits carried out using Earth remote sensing materials, the border of oil potential in the South of Western Siberia has shifted to the latitude of the urban area Lebyazhye of the Kurgan region.

The analysis of the spatial distribution of areas (grounds) with signs of the Kalchinsky, Polunyakhsky, Taytimsky and Tevrizsky oil and gas and gas condensate fields, identified by remote sensing data, and Graben-shaped hollows identified by regional seismic work, showed: a) potential grounds are mainly located within hollows; b) potential areas gravitate to the side parts of the hollows; c) the greatest number of grounds is fixed on the eastern boards of hollows.

The materials of geological-geochemical and geophysical studies do not exclude the possibility of the presence of hydrocarbon deposits in the Paleozoic floor of the southern regions of Western Siberia. But in this case, when attracting remote sensing data, other signs of the ES will be taken into account, and, accordingly, another zoning of the territory, another classification and other standards. These procedures will already be aimed at finding hydrocarbon deposits in the foundation.

The results of these studies can significantly optimize oil exploration.

References

1. I.V. Shpurov, J.E. Faibusovich, V.A. Ryl'kov, Science and Energy, **1**, 10-16 (2011)
2. Y.A. Tsymbalyuk, F.Y. Borkun, Y.A. Shepelev, in the collection: NEFTGAZTEK Materials of the VI Tyumen International Innovation Forum. Government of the Tyumen region, committee on innovations of the Tyumen region, 438-449 (2015)
3. A.R. Kurchikov, A.Y. Belonosov, O.S. Martynov, Материалы VII научно-практической конференции «Ways of the Western Siberia's oil and gas potential realization of Khanty-Mansiysk Autonomous Okrug», **1**, 41-44 (Khanty-Mansiysk city, 2003)
4. A.Y. Belonosov, R.I. Timshanov, A.E. Kudryavtsev, S.A. Sheshukov, Academic Journal of Western Siberia, **12**, 3(64), 5-6 (publishing house "Vector-Buk", Tyumen, 2016)
5. A.Y. Belonosov, D.V. Borisov, A.E. Kudryavtsev, S.A. Sheshukov, Interexpo GEO-Siberia-2017. XIII International Scientific Congress «Subsoil Use. Mining engineering. Directions and technologies for the search, exploration and development of mineral deposits. Geoecology», **2**, 172 – 175 (Novosibirsk, SSUGT, 2017)
6. V.N. Voronov, Y.A. Tsymbalyuk, L.V. Smirnov, Science and Energy, **6**, 43-50 (2011)
7. N.P. Kirda, Geology and Mineral Resources of Siberia, **3**, 39-49 (2013)
8. N.P. Kirda, Mountain journal, **10 (113)**, 20-39 (2013)

9. V.M. Matusevich, A.R. Kurchikov A.V. Rylkov, News of higher education institutions. Oil and gas, **2**, 4-13 (2011)
10. V.I. Gorny, S.G. Kritsuk, Second international conference «Prospects for development and opening of fuel and energy base of the North-West economic region of the Russian Federation», Thesis of reports, 41-42, (2000)
11. A.Y. Belonosov, A.E. Kudryavtsev, D.V. Borisov, Journal of Siberian Federal University. Engineering & Technologies, **10(6)**, 758 - 763 (2017)
12. V.I. Gorny, I.V. Stepanov, Exploration and protection of subsoil, **9**, 39-43 (2001)

Natural cracking and the methods of its research when searching for oil and gas deposits (for example the southern areas of the Tyumen region)

Andrew Belonosov[1,2], and *Anton* Kudryavtsev[2]
[1]Tyumen Industrial University, 625000, Tyumen region, Tyumen, Russia
[2]West Siberian branch of the Trofimuk Institute of Petroleum Geology and Geophysics of Siberian Branch Russian Academy of Sciences (WSB IPGG SB RAS), laboratory « Geology of oil and gas», 625026, Tyumen region, Tyumen, Russia

Abstract. Thanks to the complex approach to the study of natural cracking (use of modern remote and geochemical research) in the southern regions of Western Siberia (south of theTyumen region), large regional tectonic blocks and depression zones have been identified by space and geological-geophysical data. Their potential for detecting hydrocarbon deposits was evaluated, taking into account the geodynamic and fluid dynamic approaches.

Introduction

Natural crack zones and faults of various orders are inextricably linked with oil and gas saturation of reservoirs. It has been established that within the fields there are different-scale blocks and their structures delimiting - inter-block zones. In this regard, in recent years, geodynamic and fluid dynamic approaches based on the use of space and ground-based geophysical and geochemical information have been widely used to explain oil and gas accumulation processes [1].

The history of the issue, problem state, methods for solving it

Deciphering satellite images in the visible and infrared ranges, analyzing maps by the gravitational field Δg, magnetic field ΔT and structural maps by the roof of the pre-Jurassic base, the southern part of Western Siberia (south of the Tyumen region) can be conditionally divided into large tectonic blocks - Schuchinsk, Karabash, Miass, Tobolsk, Alym, Severo-Alym, Zaozern, Alymsko-Kalchisk, Nagornensk, West Nagornensk, Abalaksk, East Abalaksk, Mikhailovsk, East Mikhailovsk.

Structures are clearly distinguished between the blocks, which according to the marks of depths (heights) can be characterized as inter-block depression zones - Karabash-Noskinskaya, Sogrinsko-Kalchinskaya, Greater-Uvatskaya, Chelnokovsko-Prirakhtovaya, Malinovsko-Prirakhtovaya, Nagornensko-Tyapuginskaya, Abalaksko-Tobolsk, Abalaksko-Malinovskaya, Karabash-Sogrinskaya.

Thus, in the regional plan, the southern part of the Tyumen region has a fault-block structure. Along the depression zones there are signs of horizontal displacement of large tectonic blocks. Depression zones have a northwest and northeast direction and extend quite a long distance - from a few hundred to the first thousand kilometers. According to seismic data, depressions of an isometric or ellipsoidal shape are noted at the intersections of the depression zones.

Over the past 25 years, it has been shown that large domed structures, to which hydrocarbon deposits are associated, are, as a rule, also "broken up" by tectonic disturbances in the northwest and northeast directions. Moreover, the anticlinal structures themselves and the surrounding area are complicated by small overlap faults, grabens and horsts. Thus, only at a different hierarchical level, the inheritance of crack systems from large tectonic blocks and depression zones to smaller, local ones is noted.

On the numerous structural plans of the roof of the pre-Jurassic foundation, systems of northwestern and northeastern tectonic disturbances are clearly traced everywhere in many territories. Therefore, it can be stated with confidence that these crack systems were laid back in the Paleozoic time and are "operating" to the present moment.

In addition to the above-mentioned system of tectonic disturbances, the meridional-latitudinal lineament system was recorded on the structural plans of the Upper Jurassic, Cretaceous, Paleogene-Neogene and Quaternary sediments, up to the very day surface, which subsequently found its reflection on temporary seismic sections. Probably, this system arose in the Late Cretaceous period of tectonic activation that occurred in Western Siberia. Numerous deposits of gas, gas condensate and oil are confined and controlled by this particular system.

At the regional level, the meridional-latitudinal crack system is reflected in the form of the Mendeleev depression zone, which belongs to the Ural-Kazakh edge deflection.

For example, consider the intersection of two regional depression zones: the Greater-Uvat (north-eastern direction) and the Abalak-Malinov (north-western direction). Junction node of these two depressions an elongated depression in the north-west direction depression called the Abalak-Uvat basin.

In connection with the discovery of large and giant deposits in the Shirotniy Priobye (Samotlor, Pokachevskoye, Fedorovskoye, Surgutskoye, Pravdinskoye, Salymskoye, Priobskoye, Krasnoleninskoye, Zapadno-Talinskoye, etc.), the southern regions of the Tyumen region for a long time remained in third roles.

However, in the 80-90s of the last century A.L. Klopov [2-6] conducted oil prosspective research in the south of the Tyumen region using the geo-indication method. This made it possible to create maps of oil-promising cosmophotomalanias. The ideological part of the method was that hydrocarbons seeping from the hydrocarbon deposits into the overlying sedimentary deposits saturated the upper part of the soil and plants. As a result of this, the previous changed the spectral brightness coefficient and corresponded to a different phototone in cosmophotographic images.

On the northeastern, eastern, and southern sides of the Abalak-Uvat basin, extensive promising cosmophoto anomalies (CFA), presumably representing oil-bearing lands, are distinguished. Southern CFA is spatially located in the northern part of the large Mikhailovsk tectonic block. Earlier, within this block, the Mikhailovskaya 1 exploration well was drilled, and within the Abalak-Malinovsk depression, closer to its southwestern side, respectively, the Mikhailovskaya 2 exploration well. When testing the supposedly promising intervals of gas and oil inflows were not obtained. As shown by geo-indication studies, these wells were drilled outside the prospective CFA, in a phototone that displays unpromising lands.

Later, at the beginning of the XXI century, according to the method of qualitative and quantitative analysis of space materials using the reference classification for predicting oil prospective areas [7–9] in the northern part of the biggest Mikhailovsk tectonic block, as well as within the Abalak-Malinov depression zone, several small , medium and large anomalies, which according to the set of features corresponded to the Taylakovsky (east Khanty-Mansi Autonomous Okrug), Yakkun-Yakhsky (south Khanty-Mansi Autonomous Okrug) and Polunyakhsky (south Khanty-Mansi Autonomous Okrug) oil fields.

As a rule, the results of distance research are subject to ground verification (verification). Employees of the West Siberian branch of the IPGG SB RAS in the northern part of the Mikhailovsk tectonic block were carried out ground-based complex geochemical research. Their goal was: confirmation of the filtration mechanism of the formation of geochemical anomalies in zones of tectonic disturbances; confirmation of promising cosmophoto anomalies identified by the geo-indication method; confirmation of oil promising areas identified by the method of qualitative and quantitative analysis of space materials using the reference classification and identification of the most promising areas in order to search for hydrocarbon deposits.

Before conducting ground-based geochemical work, cosmogeological research were carried out.

They are aimed at identifying and studying the structures of the upper part of the sedimentary cover, promising for the migration and accumulation of hydrocarbons.

Cosmogeological research methods can be conditionally divided into: cartographic, structural-geomorphological and lineament. Cartographic are associated with the construction of special maps and geodynamic content schemes that allow you to determine the plotting structures favorable for the formation of oil and gas fields. Structural-geomorphological methods are effectively used in the direct search for hydrocarbon deposits. Lineament methods are based on the analysis of dislocations and deformations of the upper part of the sedimentary cover and, recently, have been successfully adapted to the search, exploration and exploitation of oil and gas deposits.

The results of cosmogeological research of different scales make it possible to trace tectonogenic objects that appear on a day surface in the form of various structural and morphological features of the research territory [10-14].

The methodology for processing and interpretation of cosmogeological data consists of three parts: processing and interpretation of satellite imagery materials in various ranges (visible, infrared, etc.), geological interpretation of seismic, gravimagnetic, thermal and other geological and geophysical (GIS, petrophysics, etc.) data and integrated geological interpretation of Earth remote sensing materials.

Using this methodology, systems of geodynamically stressed zones (GDSZ) and tectonic blocks were identified. GDSZ systems are weakened linear sections of the earth's surface. Conventionally, they are vertical channels for the translation of deep fluids in the form of a narrow parallelepiped. Projection GDSZ from the earth's surface deep into the section, with their confirmation on temporary seismic sections, are treated as faults of the sedimentary cover.

One of the oil-search attributes at the analysis of the study area is the presence of the weakened in the tectonic relation place of crossing (junction) of the GDSZ. It is in such places that local geophysical, geochemical, hydrogeological and other anomalies are recorded.

This approach is the main and most effective tool for predicting and diagnosing the oil and gas potential of the subsoil of the research territory.

The final stage of the cosmogeological methodology is the construction of a model of the study area in the fault-block variant.

Results

The result of cosmogeological research on Mikhailovsk area was the identification of three systems GDSZ: 6 - north-west, 5 - north-east and 2 - meridional directions.

By comparison the results of a qualitative and quantitative analysis of thermal cosmic images with GDSZ systems, it was found that 2 large and 2 small promising anomalies within the Mikhailovsky tectonic block are confined to nodes of geodynamically stressed zones. In addition, 2 large promising anomalies within the Abalak-Malinov depression zone are located on the on a northeast board and also coincide with the nodes of GDSZ.

The results of ground-based complex geochemical research (verification) on Mikhailovsk area showed that:

1. Abnormal concentrations of mercury emanations from ground samples from a depth of 2 meters gravitate, on the one hand, to the GDSZ, forming elongated linear anomalies,

and on the other hand, to the nodes of the GDSZ. Moreover, linear anomalies are confined to the structural nose of the Mikhailovsky anticline structure (within the tectonic block), and the anomalies in the nodes of the GDZ are located on the southwestern side and in the middle part (on the saddle between the Iksky deflection and the Ivanovo depression) of the Abalak-Mikhailovsk depression zone.

2. The oil currently extracted conditionally consists of 70% alkanes (methane homologs) and 30% arenes (benzene homologs). The total concentrations of vaporous alkanes (hexane, heptane, octane, nonane and decane) and arenes (benzene, toluene, xylenes) in ground samples from a depth of 2 meters have a slightly different distribution compared to the area distribution of mercury emanations. On the one hand, the chains of anomalies of these hydrocarbon components gravitate toward GDSZ the north-west directional, including nodes, and on the other hand, the most contrasting anomalies are noted on the sides of the Abalak-Mikhailovsk depression zone, as well as in the northern part of the saddle between Iksky deflection and Ivanovo depression. Within the Mikhailovsk anticlinal structure, anomalies of alkanes and arenas were not recorded. From this it becomes clear why the Mikhailovskaya 1 prospecting well, drilled in the dome of the structure, turned out to be unproductive.

3. The distribution of the activity parameter of hydrocarbon-oxidizing bacteria (HOB), which are "fed" by hydrocarbon emanations, practically repeats the areal location of anomalies of alkanes and arenes. This confirms not only the cross-cutting ("open", fluid-conducting) nature of the identified tectonic disturbances, but also the long process of vertical migration of hydrocarbon components from a probable deposit to the day surface.

4. Spatial coincidence of anomalies in alkanes, arenas, and parameter activity HOB, as a rule, is a reliable sign for detection of hydrocarbon deposits.

5. The results of 2D seismic surveys carried out later confirmed the correct orientation of the identified GDSZ and the presence of hydrocarbon traps in the section beneath the recorded geochemical anomalies in the Triassic deposits (within the Mikhailovsk tectonic block) and in the Lower Cretaceous and Upper Jurassic deposits (within the Abalak-Malinov depression zone).

6. Geochemical research carried out before, during and after seismic surveys on Mikhailovsk area revealed the fact that, when seismic impacts on the geological environment, part of the GDSZ (tectonic disturbances) are revealed (become fluid-conducting), and the other part is "sealed", i.e. remain impenetrable.

Conclusion

Thus, research on the identification of natural cracking at various hierarchical levels showed that in the northern part of the Mikhailovsk tectonic block deposits of the pre-Jurassic base are promising , and in within the Abalak-Malin depression zone - deposits of a sedimentary cover.

References

1. A.Y. Prokhorov, A.R. Kurchikov, O.S. Martynov, A.D. Mitrofanov, A.A. Korobeinikov, E.Y. Tryachin, Scientific and technical collection News of gas science, **9**, 158-166 (2012)
2. A.L. Klopov, Ways to improve the efficiency of exploration for oil and gas in the Tyumen region (Pr. of WSOGRI), **148**, 57-61 (Тюмень, 1979)
3. V.V. Borovsky, A.L. Klopov, L.L. Podsosova, I.D. Peskovsky, Exploration of the Earth from Space, **6**, 80-86 (Science, Moscow, 1980)
4. A.L. Klopov, Bulletin of the subsurface user, **2**, 105-107, (Khanty-Mansiysk city, 1999)
5. A.L. Klopov, M.V. Larina, Bulletin of the subsurface user of Khanty-Mansi Autonomous Okrug, **24** (2011)

6. A.L. Klopov, Bulletin of the subsurface user of Khanty-Mansi Autonomous Okrug, **25** (2012)
7. V.I. Gorny, B.V. Shilin, G.I. Yasinsky, Thermal aerospace survey, 128, (Nedra, Moscow, 1993)
8. V.I. Gorny, S.G. Kritsuk, Second international conference "Prospects for development and development of the fuel and energy base of the North-Western economic region of the Russian Federation", Abstracts, 41-42 (2000)
9. V.I. Gorny, I.V. Stepanov, Exploration and protection of mineral resources, **9**, 39-43 (2001)
10. A.Y. Belonosov, A.R. Kurchikov, S.V. Prozorov, Materials of the II All-Russian scientific conference "Geology and oil and gas potential of the West Siberian megabasin", 78-83 (TSOGU, Tyumen, 2001)
11. A.R. Kurchikov, A.Y. Belonosov, O.S. Martynov, Materials of the VII scientific-practical conference «Ways of the Western Siberia's oil and gas potential realization of Khanty-Mansiysk Autonomous Okrug», **1**, 41-44 (Khanty-Mansiysk city, 2003)
12. A.Y. Belonosov, O.S. Martynov, S.A. Sheshukov, International Academic Conference "State, Trends and Problems of the Development of Oil and Gas Potential of Western Siberia", Conference proceedings (Tyumen, 2008)
13. A.R. Kurchikov, A.Y. Belonosov, A.S. Fefelov et al., Materials of the All-Russian Scientific and Practical Geological Conference "Geological Survey and Mineral Resources of Russia on the Threshold of the 21st Century", 206-209 (VNIGRI, St. Petersburg, 2000)
14. A.R. Kurchikov, A.Y. Belonosov, Bulletin of the subsurface user of Khanty-Mansi Autonomous Okrug, **27**, 58-60, (Tyumen, 2001)

Methods of calculating landslide volume using remote sensing data

Maira Razakova [1], *Alexandr* Kuzmin [1], *Igor* Fedorov [1], *Rustam* Yergaliev [1] *and Zharas* Ainakulov [2,*]

[1]JSC "National Center for Space Research and Technology", 050010, Almaty, Kazakhstan
[2] Al-Farabi Kazakh National University, 050040, Almaty, Kazakhstan

Abstract. The paper considers the issues of calculating the volume of the landslide from remote sensing data. The main methods of obtaining information during research are field observations. The most important results of field studies are quantitative estimates, such as the volume of the embankment resulting from a landslide, morphometric indicators, etc. The study of a remote and remote object was carried out by remote methods using aerial photographs in the Ile Alatau foothills at 1,600 meters above sea level. The obtained materials from the mudflow survey will be useful in developing solutions to mitigate the effects of disasters and in the design of measures for engineering protection from landslides.

1 Introduction

The consequences of the mudflow can be catastrophic - the destruction of roads, industrial facilities and settlements. The obtained materials for determining the volume of mudflow will be useful in the development of solutions for the elimination of the consequences of disasters and in the design of measures for engineering protection from mudflows.

The aim of this work was to assess the environmental impacts of mudflow according to aerial photographs.

The object of the study was the events of May 14, 2018, when in the Aksai gorge due to heavy rainfall in the form of rain, a mudflow of 800 cubic meters occurred along the Kyzylzhar river (a tributary of the Aksai river) [1].

In practice, mudflow characteristics are calculated according to the requirements of [2-3]. One of the main parameters of mudflow, affecting the choice of measures, is the volume of mudflow V_c in m^3 determined by the formula:

$$V_c = V_k + V_t + V_p + V_{pw} \qquad (1)$$

where V_k is the volume of coarse material deposited in the removal cone, m^3; V_t is the volume of coarse material deposited in the transit zone, m^3; V_p is the volume of finely

[*] Corresponding author: jaras1987@mail.ru

dispersed material carried out by the mud flood beyond the boundaries of the mud flow, m^3; V_{pw} is the volume of the water after-debris flood, m^3.

The volume of mudflow of any type in the presence of observations of mudflows can be approximately determined by the formula:

$$V_c = 0{,}5 \cdot Q_{max} \cdot T \tag{2}$$

where Q_{max} is the maximum mudflow rate, m^3 / s; T– duration of mudflow, c.

To identify the general patterns of the emergence and development of catastrophic mudflows and landslides, a comprehensive detailed study of the entire territory covered by the process is necessary.

The main methods for obtaining information during research are field observations. The most important results of field studies are quantitative estimates, such as the volumes of mudflows, morphometric indicators, etc. Studying a remote and inaccessible object was carried out by remote sensing methods using aerial photographs in the foothills of Ile Alatau at an altitude of 1600 meters above sea level. The prevailing volumes of mudflows are usually formed due to moraine and alluvial-proluvial accumulations [4].

In this paper, the authors proposed an alternative method for calculating V_c, V_k, V_t, V_p based on the construction of a digital terrain model (DTM) based on aerial photography.

Aerial survey of the study area was carried out by an unmanned aerial vehicle (UAV), hexacopter [5-6]. Field topographic and geodetic works were preliminarily carried out for the placement of identification marks on the terrain, on which planned high-altitude measurements were made. Areas of accumulation of mudflow deposits are identified, heterogeneity of mudflow structure is classified, in our case, mud-stone mudflow spread over several kilometers prevailed, a microrelief of the mudflow channel surface was built. Mudflow deposits were insignificant in the form of mud-stone mass along the river bed. On the territory of the mudflow, the bulk of the mudflow occurred in a landslide. The calculated volume of the embankment formed as a result of the landslide occurred, according to the digital terrain model, coincided with ground-based measurements.

To identify the structural and topographic features of the landslide processes, we used photogrammetry methods, geoinformation technologies for constructing elevation contours, slopes, volume calculations, etc.

A lot of work has been done on the information and analytical support of such a natural phenomenon as a landslide.

2 Research methods

One of the stages of the recognition and fixation of a landslide was space monitoring. In order to study the cause and effect parameters of the landslide occurrence process that are characteristic of a given area, first of all, remote sensing (RS) data were used. At the initial stage, remote sensing was analyzed to detect changes in the earth's surface, or characteristic signs of landslide processes, such as soil slipping, soil exposure [7].

A general overview of the terrain was made and a survey was made with the UAV camera Sony NEX-5N of the site where the mudflow passed. The survey was carried out from a height of 200m above the level of the surface (soil). The classification of altitude data is based on the ranking of colors from blue to red. The high spatial resolution of the DTM allows us to talk about the microrelief - local relief forms such as embankments, sediments, etc. are distinguished. By traditional methods for calculating the characteristics of mudstone mudflows of glacial origin, the mudflow channel (from the largest glacier to the settlement section) is divided into 10 - 20 characteristic sections so that about half of them are at angles over 6°. The length of the calculation section can vary from 300 to 2000 m. The calculation is made sequentially in sections [3].

3 Results

Based on the survey materials, an orthophotomap was constructed in the visible wavelength spectrum and a DTM was obtained. The spatial resolution of the orthomosaic is 5 cm and the DTM is 12 cm (Fig. 1).

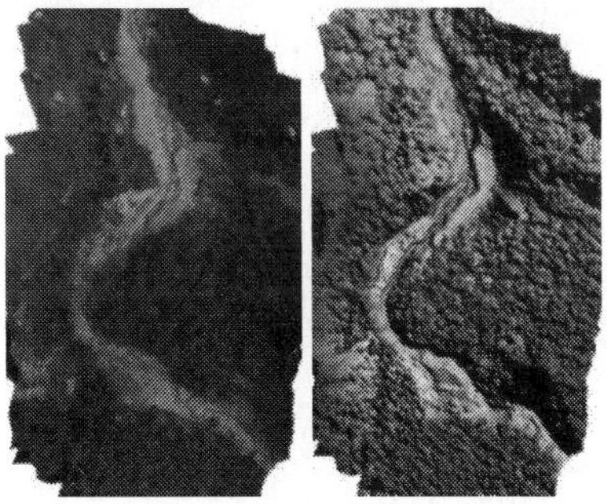

Fig. 1. Orthophotomap and DTM in the color palette according to the data from the UAV survey, the territory where the landslide occurred on May 14, 2018.

In the place where the landslide occurred, an embankment formed, the contour is highlighted in red in Fig. 2.

Fig. 2. 3D image of a DTM fragment obtained from aerial photography.

To calculate the volume of the embankment, a surface was constructed from the ground points of the contour of the embankment and the height surface of the embankment was selected (Fig. 3).

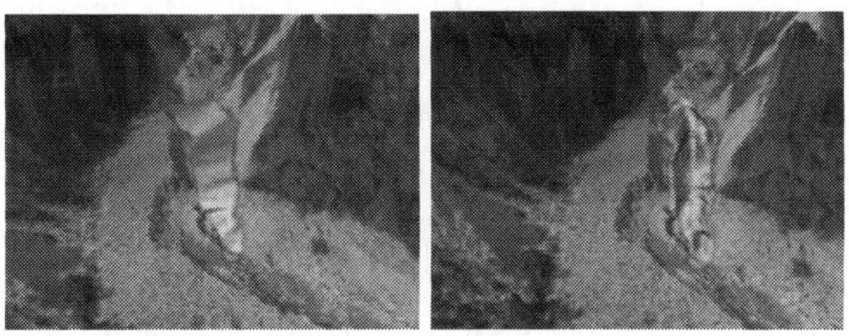

Fig. 3. 3D image of a DTM fragment, with a surface along the ground points of the embankment contour and a height file with the selected embankment surface.

As a result, a volumetric representation of the body of the embankment was obtained. (Fig. 4).

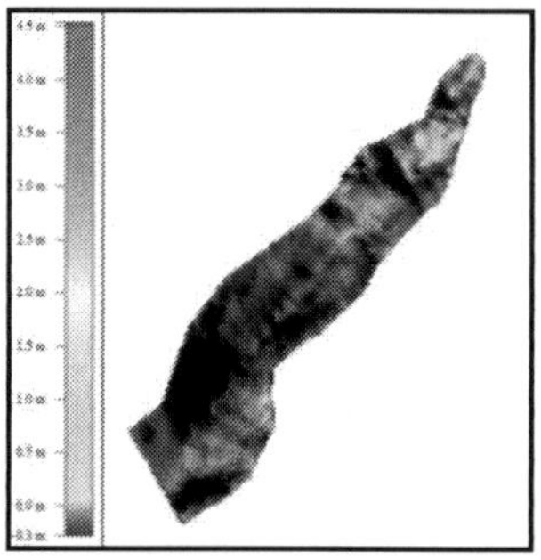

Fig. 4. Height file for the selected mound surface.

The volume of the embankment was calculated by geographic information methods, summing up the pixel-by-pixel components of a three-dimensional figure and amounted to 713.5 m3 [8]. The calculated volume according to aerial survey and ground-based measurements [9] coincided.

4 Conclusion

Since mudflow landslides threaten nearby settlements, it is necessary to carry out their constant monitoring in seasonally wet periods. Using aerial photography with UAVs, one can observe the dynamics of the movement and spread of mudflow landslides. A detailed survey makes it possible in the future to issue recommendations for measures to strengthen the places where mudflows pass, to prevent their harmful effects on the places of population living. Assess damage and devastating effects on human settlements. (In mid-2018, all the structures of the nearby dacha village Koklaisay were digitized and entered into the GIS).

References

1. https://tengrinews.kz/events/sel-soshel-bliz-almatyi-344188/
2. SP 33 - 101 - 2003. Determination of the main calculated hydrological characteristics. Gosstroy of Russia. - M .: FSUE TsPP. - 73 s (2004)
3. P - 814 - 84. Recommendations for the design of anti-mudflow protective structures. Hydroproject. - M .: Hydroproject. - 112 s (1985)
4. A.R. Medeu, V.P. Blagoveshchensky, T.A. Baimoldaev, T.L. Kirenskaya, B.S. Stepanov Mudflows of South-East Kazakhstan. - Almaty: Institute of Geography. T. 2. Part 2. The basics of monitoring in Ile Alatau. - 288 s (2018)
5. Zh.Zh. Ainakulov, N.G. Makarenko, T.T. Paltashev. Experience of modeling mining landscapes by using methods of intelligent monitoring. The journal "Current problems in remote sensing of the Earth from space". –Volume 15. –№7. - P. 43-50 (2018)
6. Zh.Zh. Ainakulov, N.G. Makarenko, T.T. Paltashev. Intelligent Monitoring on the Basis of a UAV for Assessing and Predicting the Condition of Objects with Complex Relief. International Journal of Applied Mathematics and Informatics. Volume 13 (2019)
7. M.G. Razakova, A.G. Kuzmin, Zh.Zh. Ainakulov, E.E. Aidarkhanov, I.O. Fedorov, R.K. Ergaliev. Methodical aspects of using aerospace data for monitoring hazardous

natural phenomena // Abstracts of International Symposium on Water and Land Resources in Central Asia. October 9 - 11, 2018. Almaty, Kazakhstan, 2018 .-- 1 p. (Organizers: GFZ - German Research Center for Geosciences, German-Kazakh University, University of Würzburg) (2018)

8. V.F. Kutyrev. Volume calculation using MapINFO Professional. Moscow. Geoprofi number 3. 2008. S.53-54 (2008)
9. I.M. Vaskov. On the methodology and methods of studying glacial mudflows and their genesis. // Debris flows: disasters, risk, forecast, protection. Materials of the IV International Conference Russia, Irkutsk - pos. Arshan (Republic of Buryatia) September 6-10, 2016 S. 28-36 (2016)

Application of two-dimensional fast Fourier transform algorithm, analog of the Cooley-Tukey algorithm, for 4k fixed format digital image of satellite data in frequency domain processing

Mikhail Noskov [1*], *Valeriy* Tutatchikov[1]
[1] Siberian Federal University, Institute of Space and Information Technology, Svobodny prospect 79, Krasnoyarsk, 660041, Russian Federation

Abstract. Currently, digital images in the format Full HD (1920 * 1080 pixels) and 4K (4096 * 3072) are widespread. This article will consider the option of processing a similar image in the frequency domain. As an example, take a snapshot of the earth's surface. The discrete Fourier transform will be computed using a two-dimensional analogue of the Cooley-Tukey algorithm and in a standard way by rows and columns. Let us compare the required number of operations and the results of a numerical experiment. Consider the examples of image filtering.

1 Introduction

Currently, digital images in the format Full HD (1920 * 1080 pixels) and 4K (4096 * 3072) are widespread. For their processing, spatial filtering methods are mainly used, since they do not require significant image size restrictions [1]. In this paper, we consider the possibility of using an alternative method - the frequency filtering of such images by calculating the two-dimensional discrete Fourier transform of the function of the brightness of the pixels of the image, applying a filter, and computing the inverse Fourier transform [2].

Consider the method using the example of a remote sensing image of the Earth in 4K format, consisting of 4096 * 3072 pixels. At that $N = 4096 = 2^{12}$, $M = 3072 = 3*2^{10}$. In this case, for the direct calculation of the two-dimensional discrete Fourier transform (DFT), $M^4 * N^4 = 9*2^{44}$ operations of complex addition and multiplication are required. One way to simplify the calculations is to use the classical algorithm for computing the two-dimensional fast Fourier transform (FFT): first, the one-dimensional FFT using the Cooley-Tukey algorithm for each row consisting of 4096 elements is calculated, and then the discrete Fourier transform for each column of 3072 elements. In this case, $18*2^{22} + 9*2^{32} = 9234*2^{22} \approx 9*2^{32}$ operations of multiplication and $72*2^{22} + 9*2^{32} = 9288*2^{22} \approx 9*2^{32}$ addition of complex numbers are required. In that event, the main restriction on the

* Corresponding author: mvnoskov@yandex.ru

application of the Cooley-Tukey algorithm comes into force. More precisely the size of the one-dimensional array *f(x)* should be degree 2 ($N = 2^s$). To simplify the calculations, you can convert the original image so that the width and height of the picture consisted of $N = 4096 = 2^{12}$ samples. For a given 4K format, you will have to add almost one fourth of all the pixels (which will introduce noticeable distortions into the image, or significantly expand the boundaries of the original image), and then apply a two-dimensional analog of the Cooley-Tukey algorithm [3-5]. In this case, $\frac{3}{4} \cdot N^2 \log_2 N = 9 \cdot 2^{24}$ operations of multiplication and $2N^2 \log_2 N = 24 \cdot 2^{24}$ addition of complex numbers are required.

2 Application of a two-dimensional analogue of the Cooley-Tukey algorithm

We will use another way of representing the original 4K format image. In [6], a modification of a two-dimensional analogue of the Cooley – Tukey algorithm for a rectangular signal with $N * M_1$ samples was considered. To calculate it, you will need $\frac{3}{8} \cdot NM_1 \log_2 N$ multiplication and addition $2 \cdot NM_1 \log_2 N$ of complex numbers.

We divide the original signal into 3 blocks with 4096*1024 samples by the multiplicity of the index: a multiple of 3 - up, the remainder of dividing by 3 is 1 - to the center, and the remainder of dividing by 3 is 2 - down. As a result, we obtain the sum of three sub-signals in which the sides are a power of two and can be applied [4]. In this case, the upper block is left unchanged, all elements of the central block are multiplied by a factor $e^{\pi i/3}$, and all elements of the lower block are multiplied by $e^{2\pi i/3}$. In this case, the total number of complex operations will be: multiplication - $\left(\frac{9}{8}\log_2 N + 2\right) \cdot NM_1 = 15{,}5 \cdot 2^{22}$ and addition - $(6\log_2 N + 2) \cdot NM_1 = 74 \cdot 2^{22}$.

The total number of operations required to calculate two-dimensional FFT in different ways can be presented in table 1.

Table 1. Comparison of the number of operations of addition and multiplication of complex numbers in the calculation of two-dimensional FFT by different algorithms

Method	Multiplication	Addition
Direct calculation of two-dimensional DFT	$9*2^{44}\approx 1{,}5*10^{14}$	$9*2^{44}\approx 1{,}5*10^{14}$
Cooley-Tukey FFT Row and Column DFT Calculation	$9*2^{32}\approx 3{,}9*10^{10}$	$9*2^{32}\approx 3{,}9*10^{10}$
An analogue of the Cooley-Tukey algorithm for a square signal	$36*2^{22}\approx 1{,}5*10^{8}$	$96*2^{22}\approx 4*10^{9}$
Cooley Tukey analog for 4K format image	$15{,}5*2^{22}\approx 6{,}5*10^{7}$	$74*2^{22}\approx 3{,}1*10^{8}$

As the table shows, the slowest way to calculate the two-dimensional Fourier transform for a 4K image (4096 * 3072 pixels) is to directly calculate the two-dimensional DFT (it is most rarely used due to the large number of operations required). Significant speedup of calculations allows us to obtain the classical way of calculating the two-dimensional Fourier transform: first, in rows (for a signal with 4096 samples) using a one-dimensional

Cooley-Tukey FFT, and then in columns using DFT (3072 samples). Converting the original image to a signal with 4096 * 4096 samples makes it possible to reduce by one or two orders of magnitude the number of operations required for complex multiplication and addition using a two-dimensional analog of the Cooley-Tukey algorithm. Dividing the original image into 3 sub-signals with 4096 * 1024 samples allows applying a two-dimensional FFT to each received sub-signal according to the analogy of the Cooley-Tukey algorithm for a square-wave signal, which further reduces the number of required complex number operations by an order of magnitude.

As a result, we find that the algorithm for calculating the two-dimensional fast Fourier transform using an analog of the Cooley-Tukey algorithm for a fixed 4K image (4096 * 3072 pixels) requires fewer multiplication and addition of complex numbers than the other methods presented. However, this algorithm does not require pre-processing of images in 4K format.

To test the modification of the algorithm, a program was written in the C ++ programming language. Testing was carried out on a computer with an AMD FX-4170 4.2 GHz processor, 16 gigabytes of RAM, the Windows 7 operating system. The result of comparing the running time in seconds of two-dimensional FFT calculation algorithms is presented in Table 2.

Table 2. The result of testing algorithms for calculating two-dimensional FFT in seconds

Algorithm	Signal size	Runtime
By rows and columns	4096 * 4096	*2.691*
An analogue of the Cooley-Tukey algorithm for a square signal	4096 * 4096	*1.637*
Cooley Tukey analog for 4K format image	4096 * 3072	*1.294*

As can be seen from table 2, the modified analog of the Cooley-Tukey algorithm for a 4K image is faster than the analog of the Cooley-Tukey algorithm with the addition of a signal to a size of 4096 * 4096 samples.

3 Image filtering

Consider the result of applying the described algorithm for frequency filtering of images (the filtering mechanism is described in more detail in [6–8]). As a test signal, we used a photograph of Krasnoyarsk and its environs from the LandSat-8 satellite dated July 4, 2016 in Figure 1 on the left. The original resolution of the image is 8081 * 8171 pixels [9-10]. The signal was converted to 4K format, namely 4096 * 3072, as one example of widespread images. The right-hand side of Figure 1 shows the result of high-frequency image filtering (for more details on the use of filtering in the frequency domain, see [6]). In this case, the contours became noticeable: the river, the boundaries of the rocky sections, and the image itself became more contrasting.

Fig. 1. Original image on the left - result of high-pass filtering on the right

Figure 2 on the left shows the original image, on the right - the result of low-pass filtering. In this case, small changes in the mountainous terrain are not so noticeable against the general background. The image has become less contrast and more uniform. Sudden transitions of image brightness disappeared, color saturation was added [11].

Fig. 2. Original image on the left - low-pass filtering result on the right

4 Conclusion

The article shows the use of a two-dimensional analogue of the Cooley-Tukey algorithm for frequency filtering of images in 4K format, requiring fewer multiplication and addition of complex numbers than other methods of processing such images. Several methods are compared, the results of a numerical experiment are given, and examples of image filtering based on a photograph of the earth's surface are shown.

References

1. R.C. Gonzalez and R.E. Woods. *Digital Image Processing,* (2012)
2. R. Blahut, *Fast algorithms for digital signal processing* (1989)
3. V.S. Tutatchikov, O.I. Kiselev, M.V. Noskov, *Pattern Recognition and Image Analysis*, **23**(3), 429-433 (2013)
4. V.S. Tutatchikov, *Proceeding of 11-th International Forum on Strategic Technology (IFOST-2016)*, 495 – 498 (2016)
5. M.V. Noskov, V.S. Tutatchikov, *Proceeding of the 5-th International Workshop on Image Mining. Theory and Applications*, 114-117 (2015)

6. M.V. Noskov, V.S. Tutatchikov, *Pattern Recognition and Image Analysis*, **25**(1), 81-83 (2015)
7. M.V. Noskov, V.S. Tutatchikov, *Pattern Recognition and Image Analysis*, **27**(1), 110-113 (2017)
8. О.И. Киселев, В.С. Тутатчиков, *III International Scientific Conference: Regional Problem of Earth Remote Sensing*, 158 – 161 (2016)
9. M.V. Noskov, V.S. Tutatchikov, M.P. Lapchik, M.I. Ragulina, *V International Scientific Conference: Regional Problem of Earth Remote Sensing*, 153 – 156 (2018)
10. M. Noskov, V. Tutatchikov, M. Lapchik, M. Ragulina, T. Yamskikh, *E3S Web of Conferences*, **75**, 01012 (2019)
11. M.V. Noskov, V.S. Tutatchikov, *Материалы VI International Scientific Conference: Regional Problem of Earth Remote Sensing*, 131 – 133 (2019)

Conceptual model of operational-analytical data marts for big data processing

Aleksey Raevich[1,*], *Boris* Dobronets[1] , *Olga* Popova[1] , *Ksenia* Raevich[1]

[1] Siberian Federal University, 660074, Krasnoyarsk, Borisova street 26, Russia

Abstract. Operational data marts that basically constitute slices of thematic narrowly-focused information are designed to provide operational access to big data sources due to consolidation and ranking of information resources based on their relevance. Unlike operational data marts dependent on the sources, analytical data marts are considered as independent data sources created by users to provide structuring of data for the tasks being solved. Therefore, the conceptual model of operational-analytical data marts allows combining the concepts of operational and analytical data marts to generate an analytical cluster that shall act as the basis for quick designing, development and implementation of data models.

1 Introduction

In modern and dynamically changing world, information, data and knowledge constitute the indisputable value and critical development factor both for a small enterprise and for a whole country.

Business analysis systems as a class of systems that implement the principles of decision support systems (DSS) focus on intelligent control of data and imply the use of a complex of technologies, software and practices aimed at achievement of the business objectives.

It should also be noted that detailed and quick analysis of input data for retrieval of tacit knowledge is possible only provided proper “understanding” of the data context [1]. Comparison of various data models demonstrates that specific units of information can be of small value as perceived by the end user. Information can be collected in an infinite number of sets and information elements that can co-exist in any number of various information kits. Such information should be contextualized in order to be brought to the level of the intellectual knowledge.

Therefore, the process of designing information structures to represent the data in order to fulfill certain tasks for business analysis systems shall consider the amount of data being processed, as well as the time required for a change of analytical model in case of business tasks being changed, including the changes caused by alteration of data structure in the source [2].

* Corresponding author: raevich.ap@yandex.ru

The suggested approach to generation of operational and analytical data marts shall provide availability of sources, promptness and structuring of data for tasks being solved.

2 Concept of operational data marts

Within the concept of operational data marts, data marts are considered as logically and physically decomposed subsets of data represented as sliced arrays of thematic narrowly-focused information oriented at the needs of the specific group of users.

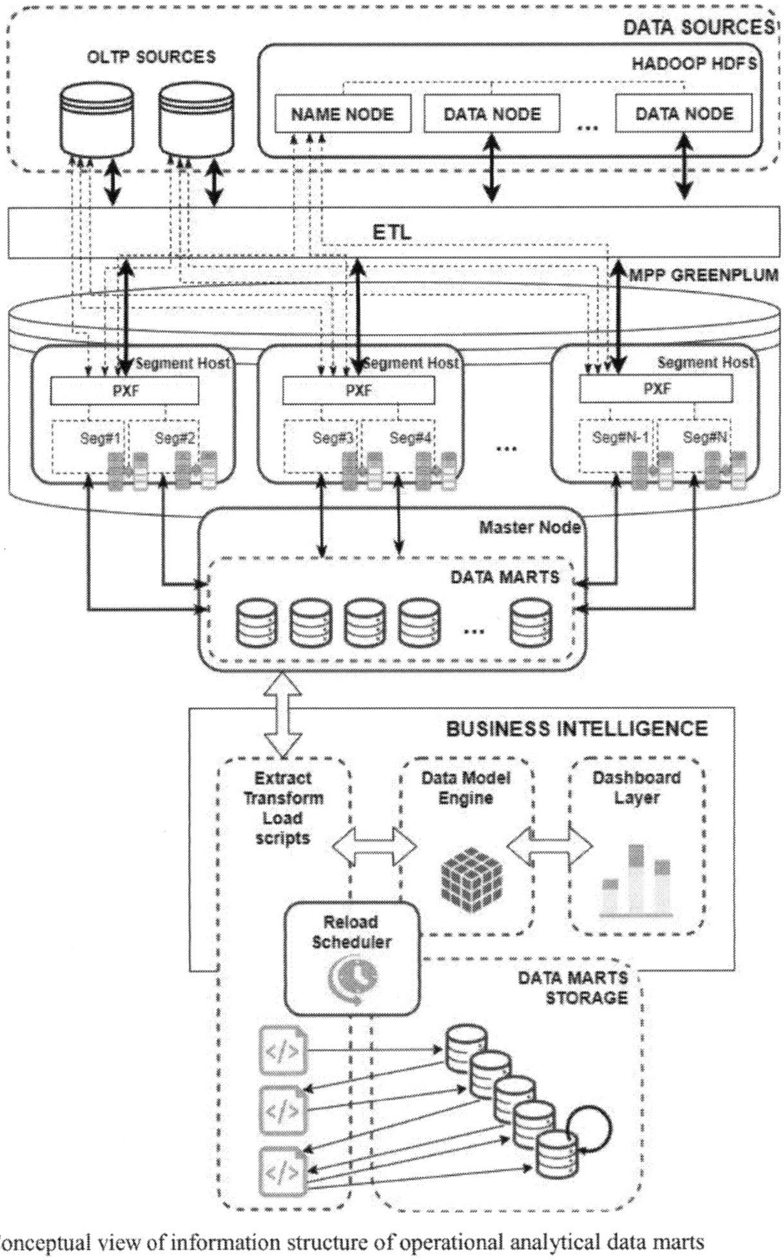

Fig. 1. Conceptual view of information structure of operational analytical data marts

In standard simulation each entity or object of the real world can be represented by a separate table with a variety of arguments. The set of jointly stored interrelated tables of minimum redundancy will be referred to as the data source.

The data source table can be designated as the information resource element a group of users shall require an operational access to. Then the information catalogue of resources shall comprise a tuple of information resources each element of which is referred to one of two classes: reference information class or detailed data class.

$$IK = \langle IR_i \rangle,\ \forall ir_{ij} \in IR_i,\ IR = \{ir_1, ir_2, \ldots, ir_N\} \tag{1}$$

It is obvious that relevance of different information resources elements based on data requests over a long time period T shall be different for different groups of users. Thus, the maximum frequency of requests for each information source element can be calculated for the preset time period:

$$f_{\max} = \max_T(\langle f_{t_1}(ir_{ij}), \ldots, f_{t_l}(ir_{ij})\rangle) \tag{2}$$

The notion of a normalized index has been introduced to rank information resources in terms of their relevance. Such index is calculated as the maximum value of the aggregate function for the tuple of information resource relevance frequency values to the frequency of user references to the information resource element within a specified period:

$$w_c = \frac{f_c}{f_{\max}} \tag{3}$$

Therefore, the current frequency of each information resource element relevance for the preset time period can be defined based on the set of values [0; 1].

Transactional systems or other storages used for big data processing, such as Hadoop can be used as information sources for operational data mart storage (Fig.1).

The most requested tables shall be uploaded to the operational data storage via filtration of information catalogue elements using the normalized index. Thus, the upload of data from the sources shall be started from the class of most relevant resources including detailed data and reference information used for detailed data contextualizing.

3 Concept of analytical data marts

Compilation of logical data models within a centralized storage is often performed with account of two main requirements: elimination of redundancy and maximum improvement of data reliability, both of which derive from approaches to collective use of data marts within the storage by a group of users.

The concept of analytical data marts basically deals with selection of profile data for the specific area of operations. And as the selection of essences, attributes and fixation of interrelations between essences depends on the semantics of the subject field and is performed by a system analyst based on his or her own subjective understanding of the application task specifics, the sets of attributes describing information objects can overlap, partially overlap or never overlap in the course of generating analytical data marts.

Business application level contains parameters converted to the terms of the specific business logic that are also projected to end user reports grouped based on business tasks as part of objects prepared and calculated at the business logic level. The process also includes calculation of field-specific parameters, hierarchy building and compilation of fact tables and measurement tables in diagrams for each business area.

Hierarchy classes of essence attributes shall be defined based on analyst needs when generating data marts for the predefined set of business processes $\{M_1,M_2,\ldots,M_n\}$. The level of data detail shall be available according to the lowest level of data detail.

For instance, for cube H_n representing the solution of a user task for business process M_i, the tuple of analytical data marts shall be defined as follows:

$$(M_i, H_{n_i}) = \langle F_{d_i}, F_{a_i}, F_{m_i} \rangle \quad (4)$$

Where F_{di} is a mart of detailed (operational) data transferred directly from data sources or from the operational storage. They correspond to elementary transactional events; F_{ai} – aggregated marts that are basically generalized values of information objects attributes; F_{mi}– metadata mart acting as an index of detailed data contents for connection of data among the marts.

Data sources shall also include any objects that contain both the structured and non-structured data that can be useful to solve analytical tasks. Analytical platform shall have access to the data directly from the source or after it has been converted to another format.

ETL layer shall extract the data from various sources and convert the data to the agreed format. The data shall be converted from one type of information to another type using data processing methods. Data processing methods shall comprise operations of data collection, data formalization and filtration, data sorting, grouping and backup, data transportation and conversion.

The internal storage of analytical data marts shall be based on the concept of QlikSense QVD BI flat indexed tables that provides for data compaction and high speed of data reading up to 100 times as compared to other data sources.

The QlikSense system core generates an associative model for data interrelation providing “on-the-go” aggregation and indexation of data via interactive transformation of the model for user needs without the need to transfer the input data from the sources.

4 Research findings

A system based on MPP GreenPlum massive parallel architecture is used as a storage for operational data marts. The architecture incorporated in GreenPlum Database software and hardware complex is based on breaking down the complete array of data into the segments that can be worked with simultaneously.

The GreenPlum architecture has been initially developed for business analytics and for analytical processing of data using standard equipment. The data segments shall be automatically distributed among several segment servers, each of which holds and controls a specific part of the total array of data.

Configuration: «segment node»: [CPU] XEON 12 cores 2.66GHz [RAM] 64GB [HDD] Hitachi scsi 3x146GB 10000rpm RAID-5. 36 segments have been configured (12 per server).

Table 1 shows the results of test export of slices arrays of thematic information from Hadoop Apache Hive source to the GreenPlum mart.

Table 1. Duration of test operations at GreenPlum cluster

Operation	Average duration, ms
Creation of an external table for hive data warehouse for test collection of data, 585,000,000 lines, 50 columns (int, float, text, datetime)	160.262
Creation of a physical table distributed randomly from the external table to 1 Gbit networks	1733700.608

Operations of joining the test sets containing 505,000,000 and 7,757,000,000 entries physically distributed based on random key	884237.118
Sorting of the generated array containing 7,700,000,000 line by one field	761334.002

It should be noted that unlike conventional DBMSs that store date strictly by lines, Greenplum can store the data being processed both by lines and by columns. This radically decreases the load on disk subsystem during statistical analysis of data. Moreover, one of the main advantages of Greenplum is the use of PXF framework that enables each segment to exchange data with sources in parallel.

Qliksense BI platform is deployed at the server of the following configuration: [CPU] XEON 24 cores 2.33GHz [RAM] 1TB [HDD] Hitachi scsi 278GB 10000rpm.

Two sets of data have been generated as test sets: #1 = 9,000,000 lines and #2 = 11,000,000 lines containing 25 and 95 columns with corresponding data types: int, float, varchar (<= 100 char), datetime. “Source 1” - local disk storage of QlikSense server. “Source 2” - GreenPlum cluster.

Table 2 shows the results of test operations for specified sources and sets of data.

Table 2. Duration of QlikSense BI test operations

Operation	**Source1, min**		**Source2, min**	
	#1	**#2**	**#1**	**#2**
Data upload without field indexing	0.23	0.27	6.8	9.88
Data upload with field indexing	0.5	0.65	7.16	10.25
Data upload with synchronization of QlikSense visual data model without field indexing	1.77	1.83	8.1	11.13
Data upload with synchronization of QlikSense visual data model with field indexing	1.8	1.85	8.2	11.1

Relevant ODBC connectors shall be used for operation with various sources in Qliksense. When working with data sources via a visual editor, any changes in upload scripts shall require update of data and synchronization with the source. For analysis of data and compilation of models in visual environment it is reasonable to save the input data in QVD internal format in a local disk storage, which shall accelerate the data reading operation up to 100 times.

5 Conclusion

When used as an operational storage of data marts, Greenplum cluster enables efficient work with big data by parallel processing of data from the sources at the cluster segments. Greenplum contains the integrated libraries of analytical algorithms with open source code that perform computations with parallel processing of data by mathematical and statistical methods and machine learning methods for structured and non-structured data.

Impossibility of operational upload of big amounts of data from sources to be analyzed in BI is compensated by use of the internal storage of analytical data marts that can be initialized by complete export of sliced arrays of data or by partial (incremental) upload.

The suggested conceptual model of operational-analytical data marts tested at the base of MPP GreenPlum and BI QlikSense software and hardware complex enables various groups of specialists to transform the marts based on big data within a short period of time (close to realtime), gain operational access to the data and consequently transform the data model for the tasks being solved within the shortest possible time. Moreover, it allows not only gain a better understanding and a deeper idea of data but also improve the data visualization for end users.

References

1. Raevich A.P., Dobronets B.S. Developing a conceptual model of operational-analytical data marts// Modelling, optimization and information technoogies. - 2019. - Vol.7 - No.4 - Pp. 1-13
2. Golov N., Rönnbäck L. Big Data normalization for massively parallel processing databases //Computer Standards & Interfaces. - 2017. - Vol. 54. - Pp. 86-93.

The Reliability of Numerical Modeling in Remote Sensing Data Analysis

Boris Dobronets[1,*] and *Olga* Popova[1,**]

[1]Institute of Space and Information Technology, Siberian Federal University, Kirenskogo 26, Krasnoyarsk, 660074 Russia

Abstract. The article deals with the problem of calculating reliable estimates of empirical distribution functions under conditions of small sample and data uncertainty. To study these issues, we develope computational probabilistic analysis as a new area in computational statistics. We propose a new approach based on random interpolation polynomials and order statistics. Arithmetic operations on probability density functions and procedures for constructing the probabilistic extensions are used.

1 Introduction

The presence of uncertainties in the remote sensing data requires the development of numerical methods that take into account these uncertainties. Thus, interval uncertainty lead to interval methods. The interval approach is actually one of the most important, but far from the only means of obtaining reliable results in mathematical computations.

Reliability can also be based on other approaches, both purely mathematical and related to computer tools for solving mathematical problems. To improve the reliability of calculations we propose to use the computational probabilistic analysis.

The paper discusses reliable estimates of remote sensing data analysis. The approach is based on the use of random interpolation polynomials and order statistics. One of the known approaches is the use of Kolmogorov-Smirnov confidence limits. Similar methods are used to construct the interval boundaries of empirical distribution functions or so called P-Boxes [1].

Information availability on the probability density function makes it possible to take into account the influence of data uncertainty in the calculations and to obtain results in the form of random variables with a constructed probability density. One approach to accounting for the random nature of the input data is Monte Carlo method [2].

With all its positive qualities, this method has several disadvantages. One of the most significant drawbacks is the low convergence rate. It is important that many practical tasks with random inputs require faster methods. Computational probabilistic analysis is one of these approaches. The main idea of computational probabilistic analysis is to use numerical operations and relations over probability densities.

In computational probabilistic analysis, various types of representations of the density function of random variables are used piecewise polynomial function. Piecewise polynomial

*e-mail: BDobronets@yandex.ru
**e-mail: OlgaArc@yandex.ru

functions are determined by grids of dimension m and the values of the functions at the grid nodes. Histograms, frequency polygons, splines, etc. are examples of such functions [4–6].

The paper considers probabilistic extensions of interpolation polynomials and splines in the case when the input data are random variables given by their probability densities.

The probability density functions of random variables x, y, z will be denoted by bold font $\boldsymbol{x}, \boldsymbol{y}, \boldsymbol{z}$. Let us identify through $\boldsymbol{R}$ the set of all probability density functions.

The interpolation problem can be formulated as follows [3]. Let f be some function, its values $f_i = f(x_i)$ at the points $a = x_0 < x_1 < x_2 ... < x_n = b$ are random variables f_i with joint probability density function $\boldsymbol{p}(f_0, f_1, \dots, f_n)$. The problem arises of approximate recovery of all realizations of the function f at an arbitrary point x.

Further, this problem will be solved be using computational probabilistic analysis and applying the concept of probabilistic extension. For these purposes, we will construct probabilistic extensions of Lagrange interpolation polynomials, piecewise linear functions and cubic splines.

2 Elements of computational probabilistic analysis

Let $(x_1, x_2, \dots, x_n)$ be a system of continuous random variables with joint probability density functions $\boldsymbol{p}(x_1, x_2, \dots, x_n)$. Let the random variable z be a function

$$z = f(x_1, x_2, \dots, x_n),$$

where $f : \mathbb{R}^n \to \mathbb{R}$.

Definition 1 *We say that the random function* $\boldsymbol{f} : \boldsymbol{R}^n \to \boldsymbol{R}$ *is a probabilistic continuation of the deterministic function* $f : \mathbb{R}^n \to \mathbb{R}$ *on the set* $D \subset \mathbb{R}^n$, *if* $\boldsymbol{f}(x) = f(x)$ *for all arguments* $x \in D$.

Definition 2 *The random function* $\boldsymbol{f} : \boldsymbol{R}^n \to \boldsymbol{R}$ *is called the probabilistic extension of the deterministic function* $f : \mathbb{R}^n \to \mathbb{R}$ *on the set* $D \subset \mathbb{R}^n$, *if*

(i) it is probabilistic continuation of f *on* D,

(ii) the probability density function $\boldsymbol{f}$ *coincides with the probability density function* $\boldsymbol{z}$ *of the random variable* $z = f(x_1, x_2, \dots, x_n)$, *where* $(x_1, x_2, \dots, x_n)$ *is a system of continuous random variables with joint probability density functions* $\boldsymbol{p}(x_1, x_2, \dots, x_n)$.

Consequently, we can write

$$\boldsymbol{z} = \boldsymbol{f}(\boldsymbol{x}_1, \dots, \boldsymbol{x}_n).$$

If at some point ξ it is necessary to directly indicate the value of the probability density function $\boldsymbol{f}$, we will use the notation

$$\boldsymbol{z}(\xi) = \boldsymbol{f}(\boldsymbol{x}_1, \dots, \boldsymbol{x}_n)(\xi).$$

The support of a continuous function f is the closure of the set $\{x \mid f(x) \neq 0\}$ and it is denoted by the symbol supp(f).

Assuming $f(x_1, x_2) = x_1 * x_2$, where $* \in \{+, -, \cdot, /\}$, we can obtain analytic formulas for determining the probability densities of the arithmetic operation results for random variables.

Let $f(x_1, \dots, x_n)$ be a rational function. We can obtained probabilistic extension $\boldsymbol{f}$ of real rational functions f by replacing (i) the real variables $x_1, x_2, \dots, x_n$ with an probability density functions $\boldsymbol{x}_1, \boldsymbol{x}_2, \dots, \boldsymbol{x}_n$ and (ii) the real arithmetic operations with corresponding probabilistic operations. The result $\boldsymbol{f}$ is called a *natural probabilistic extension* [8].

Theorem 1 ([8]) *Let $x_1, \ldots, x_n$ be independent random variables. If $f(t_1, \ldots, t_n)$ is a rational expression where each variable t_i occurs not more than once, then the natural probabilistic extension approximates a probabilistic extension.*

Theorem 2 ([9]) *Let $\boldsymbol{f}(\boldsymbol{x}_1, \boldsymbol{x}_2, \ldots, \boldsymbol{x}_n)$ be probabilistic extension of function $f(x_1, x_2, \ldots, x_n)$ and for all real t function $\boldsymbol{f}(t, \boldsymbol{x}_2, \ldots, \boldsymbol{x}_n)$ be probabilistic extension of the function $f(t, x_2, \ldots, x_n)$. Then*

$$\boldsymbol{f}(\boldsymbol{x}_1, \boldsymbol{x}_2, \ldots, \boldsymbol{x}_n)(\xi) = \int_{\mathrm{supp}(\boldsymbol{x}_1)} \boldsymbol{x}_1(t) \boldsymbol{f}(t, \boldsymbol{x}_2, \ldots, \boldsymbol{x}_n)(\xi) dt. \tag{1}$$

Corollary 1 ([9]) *Theorem 2 implies the possibility of recursive computations for the general form of probability extensions and reduction to the calculation of the one-dimensional case.*

Consider example

$$z = a_1 \boldsymbol{x}_1 + a_2 \boldsymbol{x}_2 + \ldots + a_n \boldsymbol{x}_n. \tag{2}$$

If the random variables are independent, we can calculate the value of (2) using numerical probabilistic arithmetic sequentially calculating piecewise polynomial approximations. To calculate one addition, we need Cm^2 polynomial calculations. Accordingly, the total number of calculations will be equal to Cnm^2.

Thus, the article [7] compares the number of generate random variables of Monte Carlo methods and numerical operations using probabilistic arithmetic to achieve the same accuracy. Thus, the accuracy of addition of n uniform random variables ($m = 10$) is achieved by Monte Carlo methods with the number of generate uniform random variables equal to $n \cdot 10^6$. In the case of dependent random variables, according to 1 and [9], the number of operations increases as m^n.

In the general case, Monte Carlo methods are used [2]. To overcome the shortcomings associated with the Monte Carlo methods, we will use a new approach, computational probabilistic analysis. This allows in some cases to calculate integrals of the form (1) with the required accuracy.

Next we will use random functions in the form of linear combinations

$$\boldsymbol{f}(x) = \sum_{i=1}^{n} \boldsymbol{a}_i g_i(x), \quad \text{where} \quad g_i \in C^m[a, b].$$

For this type of random functions, we introduce the following concepts. Then the formal derivative of $\boldsymbol{f}(x)$ is defined as follows:

$$\partial^k \boldsymbol{f}(x) = \sum_{i=1}^{n} \boldsymbol{a}_i g_i^{(k)}(x), \quad k = 0, \ldots, m,$$

Integral of $\boldsymbol{f}(x)$ is

$$\int_a^b \boldsymbol{f}(x) dx = \sum_{i=1}^{n} \boldsymbol{a}_i \int_a^b g_i(x) dx.$$

Function

$$f(x) = \sum_{i=1}^{n} a_i g_i(x), \quad \text{where} \quad a_i \in \mathrm{supp}(\boldsymbol{a}_i)$$

we will call *the constriction of the function* $\boldsymbol{f}$ by constants $\boldsymbol{a}_i$.

3 Interpolation problems

The interpolation problem is formulated as follows. Let the probability densities $\boldsymbol{f}_i$ be known at the points $a = x_0 < x_1 < x_2 ... < x_n = b$ and their joint probability function $\boldsymbol{p}(f_0, f_1, \dots, f_n)$ is given. We need to build a random polynomial interpolation $\boldsymbol{l}_n(x)$: $\boldsymbol{l}_n(x_i) = \boldsymbol{f}_i$.

We consider Lagrange interpolation polynomials for the case of linear interpolation. Let f_1, f_2 be known values of some function f at the points x_1, x_2. In the case of linear interpolation we obtain the exact equality

$$f(x) = l_1(x) + \frac{(x-x_1)(x-x_2)}{2} f''(\xi), \ \xi \in [x_1, x_2],$$

where l_1 is the first-degree Lagrange polynomial

$$l_1(x) = f_1 \frac{x_2 - x}{x_2 - x_1} + f_2 \frac{x - x_1}{x_2 - x_1}.$$

If the values of $f_1 \in \text{supp}(\boldsymbol{f}_1)$, $f_2 \in \text{supp}(\boldsymbol{f}_2)$ are not exactly known, it is necessary to construct a linear random function $\boldsymbol{l}(x)$ satisfying the interpolation conditions $\boldsymbol{l}(x_1) = \boldsymbol{f}_1$ and $\boldsymbol{l}(x_2) = \boldsymbol{f}_2$. Thus, using natural probabilistic extensions, we construct random Lagrange polynomials of the first degree

$$\boldsymbol{l}(x) = \boldsymbol{f}_1 \frac{x_2 - x}{x_2 - x_1} + \boldsymbol{f}_2 \frac{x - x_1}{x_2 - x_1}.$$

The interpolation function $\boldsymbol{l}(x)$ is equal to the given values at the interpolation nodes. It is important that the constriction of a random linear function with respect to the constants $\boldsymbol{f}_i$ is a real linear function. Further, if it is necessary to construct a random function $\boldsymbol{l}$ satisfying the inclusion $f \in \text{supp}(\boldsymbol{l})$, for all $x \in [x_1, x_2]$ then you will need to have a priori information about the probability density of $f'' \in \text{supp}(\boldsymbol{f}'')$ on the interval $[x_1, x_2]$. Then we can get an estimate

$$f(x) \in \text{supp}\left(\boldsymbol{l}(x) + \frac{(x-x_1)(x-x_2)}{2} \boldsymbol{f}''\right).$$

Next, we consider the general case for the Lagrange interpolations polynomial. We have

$$\boldsymbol{l}_n(x) = \sum_{i=0}^{n} \boldsymbol{f}_i \prod_{j \neq i} \frac{(x - x_j)}{(x_i - x_j)}.$$

Thus the calculation of Lagrange interpolation polynomial at an arbitrary point reduces to calculating the sum of $\boldsymbol{f}_i$ with weights. If the random variables $\boldsymbol{f}_i$ are independent the calculations are simple because they fall under the conditions of Theorem 1.

For the number of nodes $n \gtrsim 5$, the application of the Lagrange interpolation polynomials is not effective. In this case you can use piecewise linear interpolation defined by the first degree Lagrange polynomial on each interval $[x_i, x_{i+1}]$.

$$\boldsymbol{l}_1(x) = \boldsymbol{f}_i \frac{x_{i+1} - x}{x_{i+1} - x_i} + \boldsymbol{f}_{i+1} \frac{x - x_i}{x_{i+1} - x_i}. \quad (3)$$

Let us estimate the mathematical expectation of the Lagrange interpolation polynomials. In accordance with the linearity property, the expectation of the interpolation polynomial will be a linear combination of the expectations of the function values. It will coincide with the Lagrange interpolation polynomial constructed from the expectation values of the function:

$$\mathsf{E}[\boldsymbol{l}_1(x)] = \mathsf{E}[\boldsymbol{f}_i] \frac{x_{i+1} - x}{x_{i+1} - x_i} + \mathsf{E}[\boldsymbol{f}_{i+1}] \frac{x - x_i}{x_{i+1} - x_i}.$$

If, for the mathematical expectation of the random function $\boldsymbol{f}$, we are estimates of the second derivative $\max_{x \in [a,b]} |\mathsf{E}[\boldsymbol{f}^{(2)}]|$, then following theorem takes place:

Theorem 3 *[[3]] Let* $\boldsymbol{l}_1$ *be a piecewise linear interpolation of the random function* $\boldsymbol{f}$*. Then we have the estimate*

$$|\mathsf{E}[\boldsymbol{l}_1] - \mathsf{E}[\boldsymbol{f}]| \leq Kh^2 \max_{x\in[a,b]} |\mathsf{E}[\boldsymbol{f}^{(2)}]|,$$

where K is a constant independent of h.

Consider the properties of a variance for piecewise linear interpolation $\boldsymbol{l}_1$ of a random function $\boldsymbol{f}$.

Theorem 4 ([3]) *The variance of piecewise linear interpolation* $\boldsymbol{l}_1$ *satisfies following estimate*

$$\mathsf{Var}[\boldsymbol{L}_1] \leq \max_{0\leq i\leq n-1} \{\max\{\mathsf{Var}[\boldsymbol{F}_i], \mathsf{Var}[\boldsymbol{F}_{i+1}], K_{i,i+1}\}\}.$$

4 Reliable approximation of the distribution function

The section discusses the construction of reliable approximation of the empirical distribution function.

Let $x_1, \ldots, x_n$ be a sample of a random variable x with the distribution function $F(t), t \in [a, b]$. The empirical distribution function is defined as follows

$$F_n(t) = \frac{m_t}{n}, \tag{4}$$

where m_t is the number of $x_i < t$.

Consider $z_i = F(x_i), i = 1, \ldots, n$. Notice, $z_i, i = 1, \ldots, n$ are uniformly distributed random variables on $[0, 1]$. If $z_1 \leq z_2 \leq \ldots \leq z_n$, then z_k is the kth order statistic and its expectation is equal $\mathsf{E}[z_k] = k/(n + 1)$ [10].

Further, we will use the points $(x_i, i/(n + 1))$ to construct an approximation of the distribution function $F(t)$. Suppose that $\omega = \{a = x_0 < x_1 < x_2 < \ldots < x_n < b = x_{n+1}\}$ is a grid. Then we construct a piecewise linear function $l(t), t \in [a, b]$ and

$$l(x_i) = i/(n + 1), \quad i = 1, \ldots, n, \ l(a) = 0, \ l(b) = 1.$$

Note, if instead of mathematical expectations $i/(n + 1)$ if used exact values z_i, then the error of the piecewise-linear function $l(t)$ with the step $h = \max_{0\leq i\leq n-1}(x_{i+1} - x_i)$ would satisfy the estimate

$$\|F - l\|_\infty \leq Kh^2\|F^{(2)}\|_\infty.$$

Hence, the constructed piecewise-linear function fairly well approximate the distribution function F even for relatively small n.

As for z_i, we are aware that they form the order statistics. It is known that the probability density of the kth order statistic is (see [10])

$$p_k(z) = \frac{n!}{(n-k)!(k-1)!} z^{k-1}(1 - z)^{n-k}, \quad z \in [0, 1]. \tag{5}$$

The joint probability density for the vector (z_j, z_k) has the form (see also [10])

$$p_{j,k}(z_j, z_k) = \frac{n!}{(j-1)!(k-j-1)!(n-k)!} z_j^{j-1}(z_k - z_j)^{k-j-1}(1 - z_k)^{n-k}, \tag{6}$$

$$j < k, \quad 0 \leq z_j \leq z_k \leq 1.$$

For each random vector $(z_1, z_2, ..., z_n)$ we have the corresponding piecewise linear function l. Looking through all possible random vectors $(z_1, z_2, ..., z_n)$, we get the whole set of piecewise linear functions $\{l\}$. Note that $\{l\}$ contains the interpolant of the distribution function F. Hence, using the probability density of the kth order statistic for a node ξ_k, the set $\{l\}$ can be represented as a random piecewise linear function $\mathbf{l}$. Accordingly, $\mathbf{l}$ is a reliable approximation of the empirical distribution function.

5 Conclusion

We have discussed the problem of calculating reliable estimates for empirical distribution functions in a small sample. We have identified modeling and algorithmic advances necessary for success on calculation problems. For these purposes, random interpolation polynomials and order statistics are used.

We also have proposed the arithmetic operations on probability density functions and procedures for constructing probabilistic extensions as the basis of computational probabilistic analysis [3–9].

The application of the developed procedures allows us to get knowledge not only about the location of the solution area, but also to identify its probabilistic structure. There are other related issues that we can tackle: for example, the use of this approach to estimate the parameters of technical systems.

References

[1] S. Ferson, V. Kreinovich, J. Hajagos, W. Oberkampf, L. Ginzburg, Tech. Rep. 2007-0939, Sandia National Laboratories (2007).

[2] G. S. Fishman, Monte Carlo: concepts, algorithms, and applications, Springer, New York, 1996.

[3] O. Popova, Computational Technologies 22 (2) (2017) 99–114. (In Russian)

[4] B. Dobronets, O. Popova, Journal of Siberian Federal University, Mathematics and Physics 10 (1) (2017) 16–21.

[5] B. Dobronets, O. Popova, IOP Conf. Series: Journal of Physics: Conf. Series 1015 (032028). doi:10.1088/1742-6596/1015/3/032028.

[6] B. Dobronets, O. Popova, IOP Conf. Series: Materials Science and Engineering 354 (012006). doi:10.1088/1757-899X/354/1/012006.

[7] B. Dobronets, A. Krantsevich, N. Krantsevich, Siberian Federal University. Mathematics and Physics 6 (2) (2013) 168–173.

[8] B. Dobronets, O. Popova, Reliable Computing 19 (2014) 274–289.

[9] B. Dobronets, O. Popova, Tomsk State University Journal of Control and Computer Science 47 (2019) 41–48.

[10] H. David, Order Statistics, John Wiley, New York, 1981.

The objective assessment of the image quality based on the geometrical concentration

Evgenii A. *Maltsev* [1,*], Yurii A. *Maglinets* [2] , and Ruslan V. *Brezhnev* [2]

[1] Skolkovo Institute of Science and Technology, Bolshoy Boulevard 30, bld. 1, Moscow, 121205, Russia
[2] Siberian Federal University, The Institute of Space and Information Technology, Kirenskogo st. 26, Krasnoyarsk, 660074 Russia

Abstract. The method of the objective evaluation of the satellite images based on the calculation of the geometrical concentration using triangulation Delaunay is proposed. Such assessment allows to estimate the degree of the image distortion and can be used for indexing and filtration data in the satellite images catalogues.

1 Introduction

The important stage of the remote sensing data processing is the search and selection of initial data with specific requirements. One of such requirement is minimal number of the noise defects, which distorted the satellite image. Photometric distortions of space images can be of different nature: imaging equipment interference, the interference associated with atmospheric optics, noises in the receiving path of remote sensors. An example of such distortions is cloudiness cover for industrial, social and agriculture studied objects.

In the first case, the distortions can be eliminated by radiometric and atmosphere corrections, in the second case the specialist have to taking into account the cloudiness during selection of initial data.

The volumes of remote sensing data rises year by year in the regional and federal satellite images warehouses. It requires the development new criteria and metrics for searching and filtration huge volumes of the data, with using not only time-spatial properties, but using the level of relational structures of images with semantics and relationships between objects.

This paper contains the new metric of the satellite image quality based on the geometrical concentration calculation of the segmented objects on the example of cloudiness.

* Corresponding author: evgenii.maltsev@gmail.com

2 The measurement of the image quality

The subjective expert assessments are used for estimation of the image quality and for construction of the objective assessments based on the average scores (mean opinion score - MOS) [1]. However, today, the objective assessments without experts are most interest, such as:

- Peak Signal to Noise Ratio (PSNR) [2] defined with using Mean Square Error (MSE).
- Minkowski distance [3].
- Structural SIMilarity (SSIM) [4].

The described above metrics based on comparing of the 2 images: initial and distorted, but such assessments do not reflect the physiology of human perception. On the other hand, it is difficult to use such metrics because it is impossible to obtain two satellite images of the same area with the same conditions.

One of the metric of the satellite image quality is assessment of cloudiness on the image as criterion of the image distortion. The typical method is following: first stage is segmentation of image with cloudiness recognition, then the binary mask is built, and after that the cloudiness percent is calculated using following statement:

$$P = \frac{S_1}{S} 100\%, \quad (1)$$

where P – the cloudiness percent;
S_1 – the number of cloudy pixels of the binary image;
S – the total number of pixels of the binary image.

This metric of cloudiness is used in many interfaces of the remote sensing catalogs as search criterion:

- https://gptl.ru/ – Roscosmos Geoportal (Satellite images service).
- http://www.ntsomz.ru – Research Center for Earth Operative Monitoring.
- http://catalog.scanex.ru – Satellite Image Catalogue of Scanex Company.
- http://digitalatlas.ru – Satellite Image Catalogue of Institute of space and informational technologies of Siberian Federal University.

But this approach does not reflect the cloud distribution in the image [5, 6]. We propose the objective assessment of the satellite image quality without initial image based on the geometrical concentration calculation of the segmented objects on the example of cloudiness.

3 The objective assessment of the image quality based on the geometrical concentration

The algorithm of the measurement of the satellite image quality based on the cloudiness detection using geometrical concentration is described in [5,6,7], it considers the cloudiness as distorted objects.

It can be used for the construction of the objective metric as the search and filtration criterion in the satellite image catalogue. This metric will allow to estimate the image quality from the point of view of the human visual perception and its applicability for the subsequent processing of the image.

https://doi.org/10.1051/e3sconf/202014902013

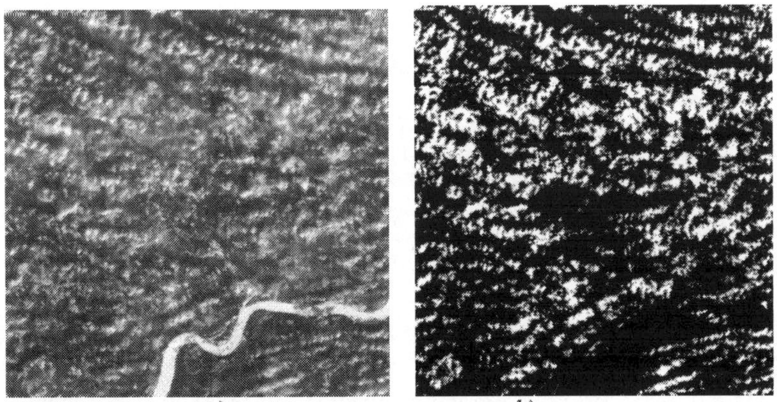

a) b)

Fig. 1. Satellite images: a) Initial satellite image with the cloudiness percent 11%; б) Binary mask of the cloudiness.

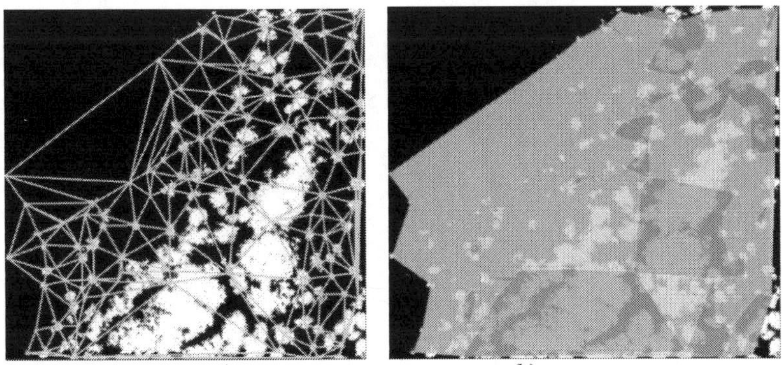

a) b)

Fig. 2. a) Calculated Delaunay triangulation (yellow color); b) Calculated geometrical concentration (low, middle, high – the gradation of green color accordingly); The binary mask of cloudiness – white color; background – black color;

The concept of calculation of the geometrical concentration and density of objects based on the using the Delaunay triangulation, also Voronoi diagram can be applied [10]. The first stage of image processing is segmentation for detection of the objects of cloudiness with using algorithm for the clouds extraction [5], with the binary mask of cloudiness creation. The next step is the Delaunay triangulation construction using binary mask of cloudiness (Fig. 2a). The next step is the geometrical concentration calculation. The coefficient of the geometrical concentration of the objects on the plane $c_i(t^i)$ [8,9] is calculated using statement (2) for the estimation of the clouds distribution in the satellite image Fig.3.

$$c_i(t^i) = \frac{\Delta NS_i}{NS_i}, \tag{2}$$

where t^i – the target class of object; ΔNS_i – the triangular area of the geometrical centers of three target objects (clouds), which are located at the minimum Euclidian distance from each other; NS_i – the sum of the objects areas.

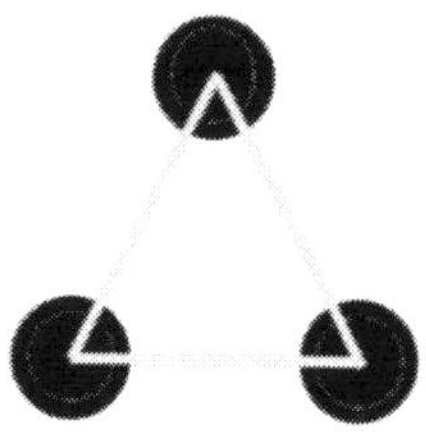

Fig. 3. Calculation of the geometrical concentration of the objects on the plane.

The measurement of triangulation elements with using (2) allows to detect of the images regions with high, middle and low geometrical concentration of objects (Fig. 2b).

The thresholds for the calculation of geometrical concentration were obtained after the images analysis of the satellite image catalogue [7]:

- 0 – 0,05 low geometrical concentration $C1$;
- 0,05 – 1 middle geometrical concentration $C2$;
- 1 – max high geometrical concentration $C3$.

The objective metric for the search and filtration of the images in the satellite image catalogue will be calculated as a ratio of a middle and high concentration area of a cloudiness to a total area of whole image:

$$K = \frac{S_{c2} + S_{c3}}{S_a} * 100\%, \quad (3)$$

where S_{c2} – middle concentration area;
S_{c3} – high concentration area;
S_a – total area of the whole image.

The calculated metric K for Fig.2 is equal = 38% but the cloudiness percent P = 23%.

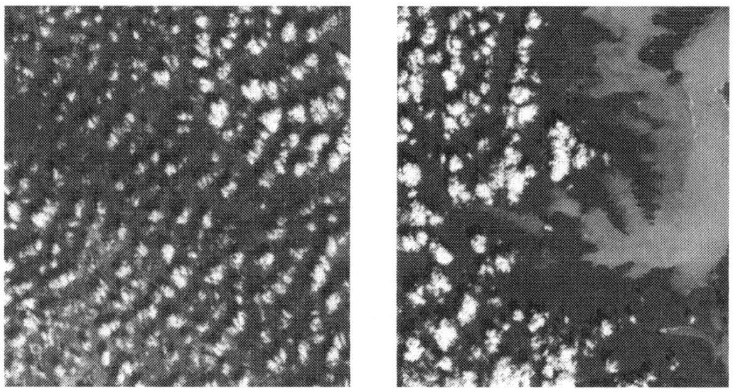

a) b)

Fig. 4. Satellite images with the cloudiness percent 11%.

The calculated metric for the image on the Fig. 4a K = 77%, but for the image on the Fig. 4b K = 43%. In spite of the equality of the cloudiness percent for the images on the Fig.4, the metric K shows the difference of the image distortion by cloudiness. The comparison of the K values for the Fig. 4a and Fig. 4b shows that image on the Fig. 4b is less distorted and more appropriated for the subsequent image processing.

This *K* estimation shows the degree of the image distortion by defects and can be built-in into the interface of the satellite images catalogue as an addition metric for the data filtrations.

4 Conclusion

New metric based on the geometric concentration calculation can be implemented as the search and filtration criterion in the satellite images catalogue. Such metric will allow to estimate the image quality from the point of view of the human visual perception and its applicability for the subsequent processing of the image.

The implementation this metric as the filtration criterion will simplify to selection satellite images by operators, it will lead to decline of the number of the unappropriated images with big value of metric K, on the other hand it can increase the number of target images with the big cloudiness percent but with low K value.

The reported study was funded by Russian Foundation for Basic Research (No. 18-47-242002), Government of the Krasnoyarsk Territory, Krasnoyarsk Regional Fund of Science, to the research project: «The development of technology for creating intelligent information systems of object-oriented monitoring of territories based on remote sensing data».

References

1. ITU-T. Recommendation P.800.1. *Methods for objective and subjective assessment of speech and video quality. Mean opinion score (MOS)* terminology, (2006).
2. R.C. Gonzalez; R.E. Woods. *Digital Image Processing*. Pearson Education, New Jersey. (2008)
3. S.Bandyopadhyay, S. Saha. *Similarity Measures. In: Unsupervised Classification*. Springer, Berlin, Heidelberg. pp 59-73 (2013)
4. Z. Wang, A.C. Bovik, H.R. Sheikh, E.P. Simoncelli. Image quality assessment: from error visibility to structural similarity. IEEE Transactions on Image Processing. 2,no.4. pp. 600 – 612. (2004)
5. E.A. Maltsev, E.E. Sirotin, D.A. Perfil'ev, G.M. Tsibulskii. *Extracting a Cluster of Clouds in the SPOT Satellite Images* // Pattern Recognition and Image Analysis. Advances in Mathematical Theory and Applications. Vol. **21**, №. 2. pp. 501–504. (2011) DOI: 10.1134/S1054661811020714.
6. E.A. Maltsev, E.E. Sirotin, D.A. Perfil'ev, G.M. Tsibulskii *The Cloud Cover Measurement of the Spot-4 Satellite Images*. Journal of Siberian Federal University. Engineering & Technologies.no.2,vol.**5**.pp.229-242 (2011)
7. E.A. Maltsev. *Development of algorithms of remote sensing data processing for recognition of agriculture spatial objects with linear geometry structure*. PhD thesis. 05.13.17 / E.A. Maltsev. Krasnoyarsk (2017).
8. D.A Perfil'ev, G. M. Tsibulskii. *Segments concentration area of metallographic images of distorted aluminum alloys*. The journal "Vestnik of SibGAU". vol. **6**. pp. 30 – 32.(2007)
9. D.A. Perfil'ev, Y.A. Maglinets, G.M. Tsibulskii *Family of Models for Describing One Class of Metallographic Images*. Pattern Recognition and Image Analysis. Advances in Mathematical Theory and Applications. vol.**19**, no.2. pp. 334-341.(2009)
10. F.P. Preparata, M.I. Shamos. Computational geometry. An introduction. Springer-Verlag, New York, (1985).

The ecological state assessment of the flora and vegetation in the parks of cities in the steppe zone of Khakasia

Elena Yu. Zhukova[1], and *Elena* G. Lagunova[1]
[1] Federal State-Funded Educational Institution of Higher Education «Katanov Khakass State University», 655017, Lenin Avenue 90, Abakan, Russia

Abstract. A study of parkland in Abakan and Chernogorsk cities, located in the steppe zone and characterized by a high level of atmospheric pollution. Plant taxa are identified that make the greatest contribution to the purification of the urban environment. The most common tree plants in the parkland are adventitious species *Populus alba* L., *P. balsamifera* L., *Acer negundo* L., *Sorbaria sorbifolia* (L.) A. Braun, *Syringa vulgaris* L., *Ulmus pumila* L. et al. In the flora of parks and squares of Abakan, 288 species of higher vascular plants belonging to 59 families and 183 genera were identified. The flora of parkland in Chernogorsk totals 142 species belonging to 33 families and 105 genera. Leading position belongs to the largest families – Asteraceae, Poaceae, Rosaceae and Fabaceae. An analysis of satellite data made it possible to remotely determine the functional state, phenological changes, and weather sensitivity of green zone vegetation. As a result of the studies, the general and seasonal indicators of the vegetation and water index were analyzed according to the satellite data (Landsat 8) for the territories of Abakan and Chernogorsk located in the steppe zone of Khakasia. Abakan had higher values of the integral NDVI (45.1-57.9), in comparison with Chernogorsk (11.4-14.2), better water supply and earlier dates of the beginning of the growing season. Based on satellite data, the productivity of green zones was compared and the regression dependences of spectral and meteorological indicators were identified.

1. Introduction

Abakan and Chernogorsk cities are located in the zone of increased natural and anthropogenic potential for atmospheric pollution, which is characterized by frequent repeatability of calm and surface inversions, which prevents the dispersion of pollutants and contributes to their accumulation. As ecological monitoring of vegetation in the urban environment is limited, the value of data on species diversity is increasing. It is of interest to compare the species composition of various cities with remote sensing, as this characterized the adaptation of species to environment, the diversity of ecological and biomorphological forms in specific circumstances. As a result, urban vegetation needs constant monitoring, for that terrestrial and satellite data are successfully applied [1-3].

The purpose of the study was to assess the current state of the flora and vegetation in the parks of cities in the steppe zone of Khakasia according to ground and satellite data.

2. Methods

Abakan and Chernogorsk are located in the Minusinsk depression, which occupies the central, lowest region at an altitude of 250-300 m above sea level. This territory belongs to the Abakan steppe district on the territory of the Uibat plain-hilly solonetzic region. The climate of the research sites is sharply continental. Atmospheric moisture is unstable and uneven. Average air temperature in July: +17.9 ° C, in January: -18.9 ° C.

Due to the synanthropization and degradation of the vegetation cover in the conditions of the studied cities, urban vegetation can be divided into cultivated (garden-park complexes, lawns) and natural (steppes, forests and meadows). Natural vegetation, preserved on the territory of cities fragmented, is slightly represented in the residential zone. Steppes, valley meadows and poplar forests are widespread in Abakan vicinity, vegetation develops under the influence of natural steppe vegetation surrounding the city and synanthropic vegetation. The hydrology of Abakan city, associated with the close location of the Yenisei, Abakan and Tasheba rivers, has a strong influence on vegetation. Chernogorsk city is surrounded by small-grained steppes. Soils in cities are poor and require land reclamation measures.

The collection of higher vascular plants in the cities of Abakan and Chernogorsk was carried out by the route method for different phenological periods from 2012 to 2018: in spring (May), summer (June-August), and autumn (early September). For the collection and herbarization of plants, a traditional technique was used. The collected plants was determined in laboratory conditions [4-5]. The floral list was compiled on the basis of our own collections, literature data and the scientific herbarium of KhSU named by N.F. Katanov. The classification by T.G. Goryshina [6], based on the ratio of plants to substrate moisture, was used for an ecological analysis of the flora. Biomorphological analysis is based on the classification by I.G. Serebryakov [7].

Landsat 8 images used in the work (http://glovis.usgs.gov/):

1. radiometric correction of Landsat 8 for 2017, cloudless images (14.05.15; 04.06.17; 20.06.17; 06.07.17; 20.07.13; 07.08.17; 20.08.16; 05.09.16) by GRASS GIS;
2. calculation of the spectral indices NDVI, NDWI [8], growth rate. To calculate the integral NDWI (Table 2), the data without correction were used (positive values), and adjusted images (negative numbers) were used for seasonal dynamics;
3. statistical analysis (mean and standard deviation), regression analysis with weather data (sum of positive temperatures and hydrothermal coefficient, HTC) (https://rp5.ru).

3. Results

288 species of higher vascular plants belonging to 59 families and 183 genera were identified in the flora of parks and squares of Abakan city. The basis of the flora is angiosperms, including 57 families (96.6%), 179 genera (97.8%), 282 species (97.9%). Vascular spore plants include 1 family of *Equisetaceae* (1.7%), 1 genus – *Equisetum* (0.6%) and 2 species (0.7%): *Equisetum arvense* L., *E. pratense* L. The gymnosperm plants include 1 family – *Pinaceae* (1.7%), 3 genera – *Larix, Picea, Pinus* (1.6%) and 4 species (1.4%): *Larix sibirica* Ledeb., *Picea obovata* Ledeb., *Picea pungens* Engelm., *Pinus sylvestris* L. Representatives of these families are background. The leading position belongs to the largest families *Asteraceae, Poaceae, Rosaceae, Fabaceae* (Table 1).

The flora of the park and squares of Chernogorsk city totals 142 species of higher vascular plants belonging to 33 families and 105 genera. The largest systematic group is the group of angiosperms in the flora of the study area. It includes 32 families (97.0% of the total number of families), 101 genera (96.2%) and 138 species (97.2%). Dicotyledonous and monocotyledonous plants are distinguished among those angiosperms. Monocotyledons were presented 3 families (9.1% of the total number of families), including 15 genera (14.3%) and 18 species (12.7%). Dicotyledons consist of 29 families (87.9%), 86 genera (81.9%) and 120 species (84.5%). Gymnosperms included one family – *Pinaceae* (3.0%), 4 genera (3.8%) and 4 species (2.8%): *Abies sibirica* Ledeb., *Larix sibirica, Picea obovata, Pinus sylvestris*.

Table 1. Family spectrum of flora of the cities of Abakan and Chernogorsk

Family	Abakan		Chernogorsk	
	number of species	share of the total number of species of flora, %	number of species	share of the total number of species of flora, %
Asteraceae	34	11.8	25	17.6
Poaceae	31	10.8	16	11.3
Rosaceae	31	10.8	15	10.6
Fabaceae	21	7.3	15	10.6
Brassicaceae	16	5.5	12	8.5
Ranunculaceae	11	3.8	6	4.2
Lamiaceae	10	3.5	8	5.6
Cyperaceae	9	3.1	1	0.7
Boraginaceae	8	2.8	4	2.8
Caryophyllaceae	8	2.8	6	4.2
Всего:	179	62.2	108	76.1

The ten leading families in the Abakan city flora account for 62.2% of the species composition of the parkland flora and 76.1% of the Chernogorsk city flora.

Leading position belongs to the largest families *Asteraceae*, *Poaceae*, *Rosaceae*, *Fabaceae.* The role of the *Brassicaceae* family is also great, the abundance of which is determined mainly by the anthropogenic factor. A high proportion of this family is a diagnostic sign of the urban flora [9].

Species distributed by humans were *Capsella bursa-pastoris* (L.) Medik., *Berteroa incana* (L.) DC, *Descurainia sophia* (L.) Webb ex Prantl, *Lepidium ruderale* L. et al. The high position in the family spectrum of the *Rosaceae* family is explained by the presence of numerous tree-shrub adventive species, which are widely used in urban landscaping. Representatives of the *Rosaceae* family are: *Pyrus ussuriensis* Maxim., *Rosa majalis* Herrm.,*Padus avium* Mill., *Spiraea chamaedryfolia* L., *Sorbus sibirica* Hedl., *Sorbaria sorbifolia* (L.) A. Braun et al.

Representatives of the *Fabaceae* family are characterized by high adaptation of species to urbanization conditions (*Trifolium pratense* L., *Amoria repens* (L.) C. Presl, *Medicago sativa* L., *Onobrychis arenaria* (Kit.) DC., *Vicia cracca* L. et al.).

The *Cyperaceae* family includes 9 species in the Abakan city flora, which is 3.1% of the total number of species (*Carex acuta* L., *C. duriuscula* C.A. Mey., *C. cespitosa* L., *C. enervis* C.A. Mey., *C. dichroa* (Freyn) V. Krecz, *Eleocharis palustris* L. et al.).

This family is represented by one species in the Chernogorsk flora – *Carex duriuscula* C.A. Mey. The role of this family in the urban flora is significantly reduced, compared with the natural flora, where it is one of the leading and ranks third in the family spectrum of the flora of Khakasia [10]. This is due to the fact that species of *Cyperaceae* family are deprived of their respective habitats in the city. According to M. A. Berezutsky (1998), it is this family that can be used as an indicator of the degree of anthropogenic floristic transformation [11].

Environmental analysis showed that the majority of the flora of the studied cities is mesophytes. They number 131 species (45.5% of the total number of species) in the Abakan flora, for Chernogorsk – 39.1%. Such an indicator is quite natural, since it meets the conditions of habitats prevailing in the city. The group of mesoxerophytes which combines 75 species or 26.0% of the total number of identified species in the Abakan flora is in second place, and 63 species (31.2%) in Chernogorsk city. Plants belonging to this group grow along roads, paths, in steppe meadow plots, along the outskirts of tree stands. A group of xerophytes is significant, because it has 39 species (13.5%) in Abakan and 49 species (24.3%) in Chernogorsk. Since the city of Chernogorsk is located in the steppe zone, where there are no water sources, the number of xerophytic species is higher here than in Abakan. There are quite a lot of hygrophytes (22 species or 7.6% of the total number of species), mesogyrophytes – 17 species (6.0%), hydrophytes are less common and are represented by only 4 species (1.4%) in the conditions of Abakan. There are enough habitats in the study area that have a favorable water regime (meadows, banks of drainage channels, etc.) where favorable conditions are created for the growth of species – hygro- and hydrophytes. Naturally, mesogyrophytes (2.5%) and hygrophytes (1.5%) are represented slightly in Chernogorsk.

The predominant biomorphs are grassy polycarps. There are 165 species (57.3%) in the Abakan flora; there are 74 species (52.1%) in the Chernogorsk one. Monocarpic plants (45 species, or 15.6% in Abakan and 38 species, or 26.8% in Chernogorsk), many of which belong to weed species; ones are settled in the studied territories by humans or animals.

Urban areas need a certain level of landscaping depending on environmental conditions. Abakan parkland totaled 24.5 km^2, or 21.9% (Table 2).

Table 2. The area, the productivity and the growth rate of green areas of Abakan (2017) and Chernogorsk (2016)

Zone	Abakan				
	S, km^2	% of S city	NDVI	NDWI	growth rate *
Preobrazhenskiy park	0.9	0.8	50.4±2.1	20.4±7.1	0.014
Square «Rainbow»	0.5	0.4	45.1±2.2	16.9±6.8	0.036
Lake st.	1.3	1.1	57.9±2.5	32.4±7.4	0.275
Krylova st.	0.6	0.6	51.4±2.4	22.8±8.1	0.017
Park of culture	8.3	7.4	56.8±2.5	29.8±6.9	0.118
Hippodrome	0.4	0.3	49.7±2.4	21.8±7.1	0.114
Chernogorskiy park	1.4	1.3	48.5±2.3	20.4±7.3	-0.030
Park "Eaglet	1	0.9	50.0±2.4	23.6±9.1	0.070
Soviet st.	6	5.4	51.4±2.2	23.6±9.1	0.284
Park "Komsomolskiy"	4.2	3.7	50.4±2.3	22.1±9.9	0.095
	Chernogorsk				
City park	0.30	1.17	12.3±0.1	10.2±0.1	0.517
Square «Young tourist»	0.02	0.09	14.2±0.1	11.4±0.1	0.632
New Year's square	0.10	0.51	13.3±0.1	10.6±0.1	0.539
Square «Stream»	0.02	0.08	11.4±0.1	8.2±0.1	0.664
Square «MultiDom»	0.01	0.05	14.2±0.1	11.7±0.1	0.492
School square (№ 19)	0.07	0.30	13.8±0.1	10.8±0.1	0.589
School square (№ 20)	0.03	0.14	11.8±0.1	9.0±0.1	0.605

* – growth of NDVI during the period of maximum plant growth (from June 04 to July 06), % NDVI for 1 day

The results of a quantitative assessment of the total productivity of the vegetation of Abakan showed that the plots did not differ much among themselves, in contrast to Chernogorsk, located in more arid conditions and in the immediate vicinity of the highway and the coal mine. The most part of Abakan city parkland had the indicator (NDVI for vegetation season) of 50.4-51.4, the most productive communities (Lake St. and Park of culture had indicators of 57.9 and 56.8). The lowest indicators were at the square «Rainbow» (45.1).

The period of growth of plants in parkland depends on the structural and functional characteristics of communities. The maximum was observed on Lake st. and Soviet st. (0.275 and 0.284, respectively).

The dynamics of NDVI is presented to identify phenological changes (Fig. 1).

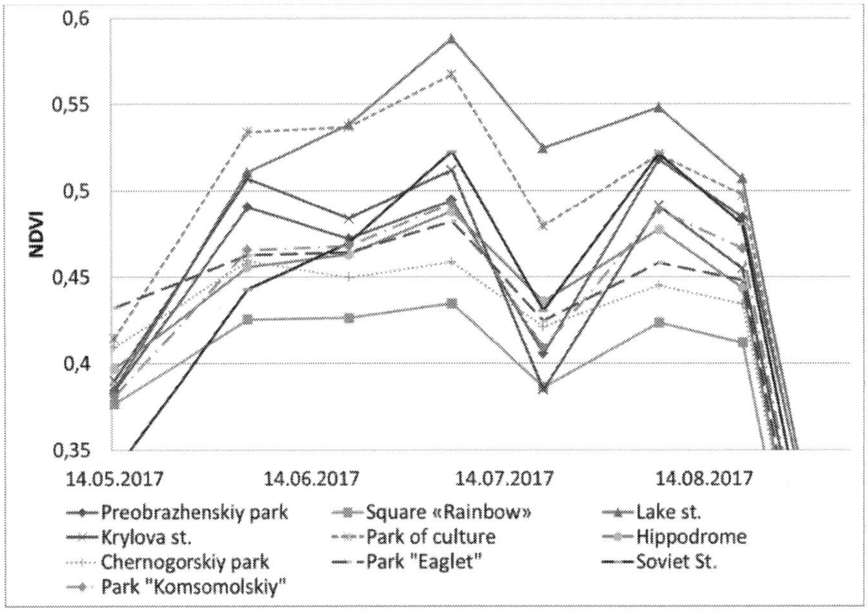

Fig. 1. Vegetation curve of NDVI green zones of the city of Abakan, 2017

All communities had similar data from 0.38 to 0.44 in mid-May 2017. The square «Rainbow» had a low NDVI throughout the season. The highest rates were observed on the Park of culture and Lake st. territories. The growth of biomass continued until the beginning of July, at which point most sites had indicators of 0.48-0.52 and clearly differed in productivity. Then, a summer break in the growing season was observed until mid-July. A secondary growth of plants (in the range 0.4-0.5) is noted by the beginning of August, one is characteristic for the steppe zone.

The water content is critical for plants located in arid or urban environments. In this regard, the seasonal variation in moisture content was revealed (Fig. 2).

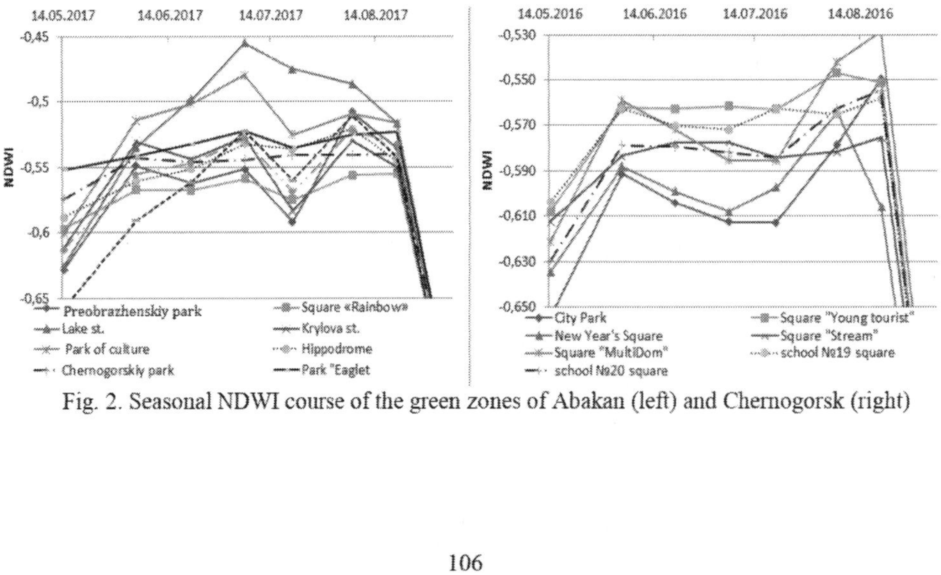

Fig. 2. Seasonal NDWI course of the green zones of Abakan (left) and Chernogorsk (right)

The course of NDWI correlates with the course of the vegetation curve (Fig. 1). It reaches its maximum values for Abakan in the middle of the season (early July) as well as NDVI, for Chernogorsk – at the end of August. Intensive plant growth can occur if there is sufficient moisture. The average daily temperature in July was higher than the annual average values (20.3 ° C versus 17.9 ° C), in June the figure was 18.6 ° C, in August 16.9 ° C in Chernogorsk (2016), 133.7 mm of precipitation fell during the summer. A favorable period was a warm and humid August (80.1 mm), as evidenced by the secondary growth of NDVI and NDWI [3]. NDWI allows you to identify communities with a good water supply: for example, by a smooth decline on the Lake st. and hippodrome (Fig. 2). There is a drainage channel providing groundwater supply in the northern part of this area. There are poplar plantings, which is drought tolerant due to the deep root system.

The highest regression indicators of satellite and meteorological data were obtained for all areas during the period of active growth from 05/14/17 to 07/06/17 for Abakan (Table 3).

Table 3. Relations of meteorological and spectral indicators, Abakan

Indicators	A regression equation	R^2
NDVI and the sum of the temperatures	$y = 0.0005x + 0.3244$	0.60
NDWI and NDVI	$y = 0.6266x - 0.8354$	0.95
NDWI and HTC	$y = 0.1938 \ln x - 0.7466$	0.74
NDVI and HTC	$y = 0.8715e^{2.45x}$	0.73

A significant dependence of the parameters is noticeable. The equation of NDVI and HTC showed a coefficient of determination of 0.73, between HTC and NDWI – 0.74.

4. Conclusion

Higher indices were observed in areas where vegetation is subject to care, watering and fertilizing, and young trees (5-8 years old) are found in the composition. The indicator for parks with trees over 20 years old is reduced. The growth rate in Chernogorsk is higher in comparison with Abakan, due to the late passage of the phenophases. Most of the investigated green areas lack moisture according to the NDWI index. Meteorological and spectral indicators showed a significant degree of positive correlation (0.60-0.74).

References

1. O. N. Vorobyev, E. A. Kurbanov, A. V. Gubayev, Yu. A. Polevshchikova, E. N. Demisheva, V. O. Koptelov, Bulletin of the povolga state technological university, **1(25)**, 5-21 (2015)
2. M. Maximov, A. Skachkova, D. Kurlovich, Land of Belarus, **4**, 42-44 (2016)
3. M. A. Leshchinskaya, E. G. Lagunova, E. Yu. Zhukova Bulletin of KrasGAU, **2(137)**, 183-188 (2018)
4. I. M. Krasnoborov et al., Key to plants of the south of the Krasnoyarsk territory (Novosibirsk, 1979)
5. *Flora of Siberia* **1-14** (Novosibirsk, 1987–2003)
6. T.K. Goryshina, *Plant Ecology* (Moscow, 1979)
7. I. G. Serebryakov, Ecological morphology of plants. Life forms of angiosperms and conifers (Moscow, 1962)
8. D. Chen, J. Huang, T. J. Jackson, Remote Sensing of Environment, **98**, 225-236 (2005)
9. A. V. Panin, M. A. Berezutsky, Botanicheskii Zhurnal. **92(8)**, 1144-1154 (2007)
10. E. S. Ankipovich, *Catalog of Flora of the Republic of Khakasia* (Barnaul, 1999)
11. M. A. Berezutsky, Botanicheskii Zhurnal. **83(9)**, 77-83 (1998)

The productivity of agrocoenosis in the steppe zone of Khakasia according to Terra Modis

Elena Zhukova[1*], *Natalya* Kutkina[2], and *Alexander* Zhukov[1]
[1] Federal State-Funded Educational Institution of Higher Education «Katanov Khakass State University», 655017, Lenin Avenue 90, Abakan, Russia
[2] Research Institute of Agricultural problems of Khakasia, 655132, Sadovaya st., 5 Zelenoe village, Russia

Abstract. The article presents the results of a quantitative assessment of the productivity of agrocoenosis according to Terra MODIS within the steppe zone of Khakasia. The gross primary production of various agrocoenosis during the growing season ranges from 3.42 to 7.41 kg/m^2. On average, the integral productivity for the growing season of 2018 was for wheat – 5.01 ± 1.04 kg / m^2, oats – 4.86 ± 1.12 kg / m^2, barley – 4.52 ± 0.42 kg / m^2, buckwheat – 5.25 kg / m^2, hayfields – 4.86 ± 0.45 kg / m^2, deposits – 4.39 ± 0.40 kg / m^2. Least of all phytomass were some arable fields and low-yielding crops of wheat and abandoned lands, and most of all perennial grasses and high-yielding wheat and oats. The GPP indicators are associated with the sum of positive temperatures positively (R^2=0.673), to a lesser extent (R^2=0.333) with a hydrothermal coefficient for the examined agrocoenosis. The most productive crops and fertile territories were determined based on the seasonal and long-term dynamics of gross productivity. An increase in productivity for some agrocenoses was noted in 2018 according to satellite data.

1. Introduction

Khakasia belongs to territories with an unstable structure of agricultural land using due to the influence of natural and anthropogenic factors, in particular the subarid climate, decreased fertility, salinization, and soil erosion and grazing. The Beisk coal deposit is located in the study area, and the coal industry is developing intensively [1]. A regular monitoring of agrocoenosis is necessary under these conditions. Remote sensing tools can significantly simplify the task of assessing the state of various plant communities throughout the growing season by ground surveys replacing or optimizing [2, 3].

The purpose of the study was to evaluate the productivity of agrocoenosis on test plots in the Beya district of Khakasia based on Terra Modis satellite data.

2. Methods

The object of the study was the soil and vegetation cover of the arable land of the hilly steppe of the Beya region of Khakasia, used for crops (agrocoenosis), hayfields – preserved arable land (fallow) and not used fallow. Agricultural land was located in the complex landscapes of a hilly plain, where moisture is redistributed due to a surface effluent, which is associated with the development of erosion processes.

The work was based on MODIS data (https://lpdaacsvc.cr.usgs.gov/appeears): 1) MOD17A2H (gross primary productivity, kg / m^2) [4]; 2) MOD16A2 (evapotranspiration)

* Corresponding author: biosara@mail.ru

[5]. Gross productivity for the season was determined by integrating the GPP curve for the period from May 1 to August 31.

Weather data provided by the site https://rp5.ru/. For regression, the calculation of the sum of positive temperatures was carried out for the period preceding the date of the survey: the sum of temperatures above 0 ° C was calculated 8 times a day (the sum of temperatures above standard). This method showed higher correlation values on the data sample.

A general route study and the establishment of decoding features of soils in satellite images were carried out. According to the main sections (depth 1.5 m), with precise reference to the terrain by GPS-navigator and digging (0.8-0.5 m), the genetic nomenclature of soils (types, subtypes, genera, species, varieties) was specified according to the classification [6]. Soil samples were taken from the genetic horizons of the soil profile for laboratory analysis and subsequent clarification of the diagnosis.

Test sites were located in the Beya vicinity (Beya district, Khakasia) on ordinary chernozems of different genera (calcareous, solonetzic, solonchakous), meadow-chernozem soils (solonchakous), as well as leached chernozems on the slopes of northern exposures adjacent to the forest-steppe (Fig. 1).

Fig. 1. Location and codes of agrocoenosis (NDVI, Sentinel, 2018): hayfields – *1к, 21 к, 22 к;* wheat – *11 к, 12 к, 14-15 к, 16 к, 18 к, 19 к, 20 к* and *10 T*; oats – *2 к, 9 T*; fallow – *3к, 6к, 7к, 8к, 9к, 10к;* arable fields – *5к, 2T, 4T, 5T, 4к, 23к, 1T, 3T;* barley – *13к, 17 к;* buckwheat – *6 T*; *Melilotus – 7 T*; *Elytrigia repens + Onobrychis arenaria – 8T*

The ones are the most fertile soils, both throughout the territory of Khakasia and on the site. Red-colored heavy loams and clays dominate among parent rocks in the study area. Leached chernozems are developed on eluvial-deluvial yellow-brown loams.

Geobotanical studies of fallows are observed in 2018 at test sites using the stationary area method [7, 8]. The productivity of agrocoenosis was determined by the standard method. Herbage was cut off at the surface of the soil on sites 1 × 1 m^2 in fivefold repetition. The phytomass of the fallows was divided into botanical and economic groups – sedges, cereals, legumes, wormwood, different herbs, rags with mort-mass. The phytomass of agrocoenosis was divided into groups – the mass of ears, vegetative mass, the mass of weeds, dry grass and mortmass. Plants were weighed in a wet and air-dry state with an accuracy of 0.01 g. Data were transferred in kg / m^2 [8, 9].

Statistical analysis (regression, calculation of the confidence interval, arithmetic mean and standard deviation) was performed in Microsoft Excel.

3. Results

The total productivity of gross productivity for the growing season of 2018 was calculated according to Terra MODIS (Table 1).

As can be seen from table 1, the most minimal value of the indicator for wheat crops (*20k*) was 3.57 kg / m^2 on meadow chernozem solonchakous, with a sulphate-soda type of salinization. The limiting yield factor in this case, obviously, is salinization, especially soda – the most toxic for most plants.

On average, the integral wheat productivity for the growing season (2018) was 5.01 ± 1.04 kg / m^2. For comparison, we present the phytomass of wheat sowing (*11 k*) in the second decade of August. The phytomass of ears was 0.091 kg / m^2 in raw form (0.061 kg / m^2 in dry), the vegetative mass was 0.084 (0.060) kg / m^2, weeds 0.231 (0.074) kg / m^2, dry grass and mortmass 0.183 (0.107) kg / m^2.

During this period, the productivity of this agrocoenosis based on MODIS data was 0.0262 kg / m^2, aboveground phytomass was significantly higher – 0.589 kg / m^2 in raw form and 0.302 in dry form. The differences are explained by the phenological state of wheat and the presence of a significant amount of dry grass, which reduces the value of the indicator.

Similar indicators were observed for barley (4.22 and 4.82 kg / m^2) and oats (4.06 and 5.65 kg / m^2). Ground-based studies have shown high grain yields in the contours: *4T*, *5T*, *9T* and *10T*. It is associated with increased moisture capacity of the heavy loamy ordinary and leached chernozems, as well as average indicators of their potential and effective fertility (humus content 4.5-5.6%, mobile phosphorus and potassium, respectively 20.5-23.9 and 214-276 mg / kg of soil). A high yield was noted according to ground data in the contours of *9T* and *10T*, which is associated with high fertility.

The barley was more productive than the wheat in accordance with ground-based data. So, the phytomass of barley ears (*13 k*) is 0.149 kg / m^2 in raw form (0.154 kg / m^2 in dry), the vegetative mass is 0.117 (0.126) kg / m^2, weeds are 0.095 (0.047) kg / m^2, dry grass and mortmass 0.135 (0.119) kg / m^2.

Weed participation (*Panicum miliaceum* ssp. *ruderale*, *Brassica napus*, *Elytrigia repens*, *Erodium stephanianum*, *Setaria viridis*, *Panicum dichotomiflorum*, *Nonea pulla*, *Oxybasis glauca*, *Thlaspi arvense*) is small. For comparison, the yield of the main agricultural crops was, according to the results of the harvesting campaign in the Beya district as of September 30 (2018): for hay – 19.1, wheat – 23.7, barley – 23.0, oats – 24.4, buckwheat – 9.1 centners per hectare.

Hayfields had values close to wheat; their projective cover did not exceed 50%. The most productive were perennial grasses of the fourth year of use – *Elytrigia repens* + *Onobrychis arenaria* and *Melilotus* (7.41 and 5.14 kg / m^2).

Table 1. Total seasonal productivity and evapotranspiration of agrocoenosis in test sites (Beya district)

The code	GPP[1]	ET[2]	R^2 [3]	R^2 [4]	The code	GPP [1]	ET[2]	R^2 [3]	Код
oats					**hayfields**				
2к	4.062	3611	0.654	0.272	*1к*	4.82	2241	0.687	0.223
9T	5.65	2505	0.632	0.188	*21к*	4.43	1710	0.701	0.472
barley					*22к*	5.32		0.737	0.389
13к	4.22	1678	0.626	0.393	**fallow**				
17к	4.82	1973	0.736	0.324	*3к*	4.56	1730	0.788	0.315
wheat					*6к*	4.09	1875	0.627	0.274
11к	4.38	1661	0.619	0.441	*7к*	4.26	2067	0.647	0.253
12к	4.22	1678	0.626	0.393	*8к*	4.10	1861	0.651	0.319
14-15 к	4.66	1963	0.558	0.296	*9к*	4.22	1678	0.626	0.393
16к	5.02	1987	0.666	0.426	*10к*	5.12	2050	0.687	0.537
18к	4.90	1820	0.737	0.389	**Elytrigia repens + Onobrychis arenaria**				
19к	4.28	1950	0.582	0.344	*8T*	7.41	3011	0.747	0.233
20к	3.57	1950	0.644	0.365	**Melilotus**				
10T	7.18	3007	0.655	0.324	7T	5.14	2072	0.704	0.455
arable fields					**buckwheat**				
5к[5]	4.87	2100	0.792	0.354	6T	5.25	2326	0.583	0.203
2T[5]	5.16	2087	0.797	0.299	**arable fields**				
4T[5]	5.65	2248	0.740	0.387	*4к*	4.56	2071	0.787	0.318
5T[5]	6.77	2888	0.743	0.267	*23к*	3.42	1665	0.514	0.264
					1T	5.21	1892	0.731	0.253
					3T	4.26	1643	0.508	0.279

[1] – gross primary productivity, kg / m^2;
[2] – evapotranspiration, kg / m^2;
[3] – coefficient of determination of the polynomial equation of the third degree of GPP correlation with the sum of temperatures;
[4] – same for GPP with hydrothermal coefficient;
[5] – cereal (wheat) was subsequently planted on these arable fields.

The productivity of the fallows varied slightly from 4.09 to 5.12 kg / m^2. For comparison, we present the ground data for the *8 k* plot (*Elytrigia repens-Vicia hirsuta*) amounted to 0.452 kg / m^2 in raw form and 0.288 kg / m^2 in dry: of which cereals – 0.121 (0.063) kg / m^2, wormwood – 0.077 (0.031) kg / m^2, legumes – 0.018 (0.008) kg / m^2, different herbs – 0.042 (0.023) kg / m^2, dry grass and mortmass – 0.193 (0.163) kg / m^2.

During the season, the phytomass of agrocoenosis intensively increased at the end of May, and then the development maximum remained until the beginning of June (Fig. 2). Further, the mass decreased except for two peaks in early July and August, associated with favorable weather conditions. The main seasonal increase in GPP was in early June, the end of the first decade of July, the end of the third decades of July.

Evapotranspiration (ET) can serve as an indirect method for assessing soil moisture content, and the GPP is higher the higher the ET (Table 1). This indicator increases with significant phytomass (due to the volume of the transpiration surface of the leaves), good water availability, and partly with a decrease in projective cover. It is worth noting that the vegetation cover to some extent prevents the drying out of the soil.

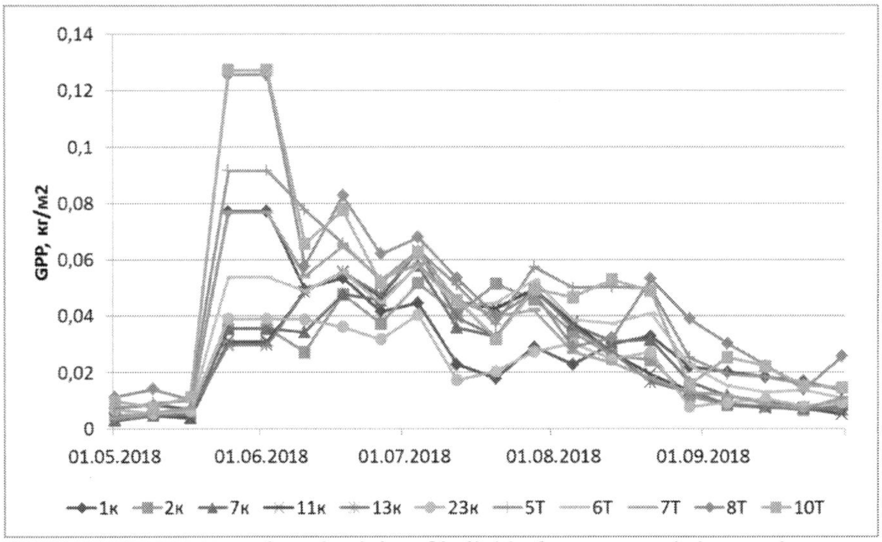

Fig. 2. Seasonal productivity of individual agrocoenosis in test plots

The relationship between weather data and distance indicators allows you to determine the degree of weather sensitivity of crops. On average, for the examined agrocoenoses, the GPP indicators are associated positively with the sum of positive temperatures ($R^2 = 0.673$), to a lesser extent ($R^2 = 0.333$) with a hydrothermal coefficient.

Long-term changes of the productivity were also analized for individual agrocoenosis (Table 2).

Table 2. Average long-term indicators of gross primary production of agrocoenosis of Khakasia (Beya district) from 2000 to 2018

agrocoenosis	$\bar{X}$ and σ	Confidence interval		GPP (2018), kg/m^2	Min	Max
		lower	upper			
1 T	4.65±0.45	4.43	4.86	5.21	3.91	5.50
2 T	4.58±0.47	4.35	4.81	5.16	3.87	5.55
3 T	4.28±0.45	4.06	4.49	4.26	3.43	5.24
4 T	4.83±0.71	4.49	5.18	5.65	3.27	5.92
5 T	5.93±0.57	5.66	6.21	6.77	5.05	7.02
Buck wheat	4.50±0.70	4.17	4.84	5.25	3.04	5.87
Melilotus	4.22±0.45	4.00	4.44	5.14	3.29	5.14
Elytrigia + Onobrychis	5.24±1.14	4.69	5.79	7.41	3.63	7.41
Oats (*9 T*)	5.02±0.47	4.79	5.24	5.65	4.20	5.96
Wheat (*10 T*)	6.04±0.55	5.77	6.30	7.18	5.07	7.18

The most productive areas were those where in 2018 were located arable fields (*5 T*), wheat (*10 T*), *Elytrigia + Onobrychis*, and oats (*9 T*). The remaining communities had lower rates. The largest positive increase in GPP among the examined sites was observed in 2003-2004, and in 2015-2016. In 2018, an increase in productivity was also noted, and even higher in comparison with 2003-2004. The productivity of the site with perennial herbs (*Elytrigia + Onobrychis, Melilotus*) is growing rapidly.

4. Conclusion

Thus, the gross productivity of agrocoenosis for the season in the study area was calculated and the differences between them were determined. Least of all phytomass were individual arable fields and low-yielding wheat crops (3.6 kg / m^2), fallows, and most of all perennial grasses and high-yielding wheat and oats (up to 5.02-7.41 kg / m^2).

Terra Modis data allows you to rank phytocoenosis by productivity and choose the optimal crops for growing in a particular area. Terra Modis satellite data showed a positive correlation with ground-based data and can be used for agroecological monitoring and optimization of land reclamation in the Beya region. The results obtained are of practical interest for the optimal placement of crops and agricultural cultivation technologies; monitoring fertility and the state of agrocoenosis; calculation of their normative productivity.

The study was supported by state contract No. F.2018.387388 (08.13.2018).

References

1. E. G. Yazikov, V. M. Khudyakov, S. V. Azarova, Bulletin of the Tomsk Polytechnic University. Geo-Resource Engineering, **307(4)**, 55-59 (2004)
2. E. Yu. Zhukova, A. P. Shevyrnogov, V. M. Zhukova, T. M. Zorkina, I. Yu. Pugacheva, Bulletin of Tomsk State University, **323**, 354-357 (2009)
3. F. V. Eroshenko, N. G. Lapenko, I. G. Storchak, Bulletin of the Orenburg State Agrarian University, **5(73)** 14-17 (2018)
4. S. Running, Q. Mu, M. Zhao, doi: 10.5067 / MODIS / MOD17A2H.006 (2015)
5. S. Running, Q. Mu, M. Zhao, doi: 10.5067 / MODIS / MOD16A2.006 (2017).
6. V. V. Egorov, E. N. Ivanova, V. M. Fridland, *Classification and soil diagnostics of the USSR* (Moscow B, 1977)
7. E. M. Lavrenko, A. A. Korchagin, *Field geobotany* (Leningrad, 1976)
8. T. M. Zorkina, *Phytocenology* (Abakan, 2003)
9. I. V. Larin, *Meadow farming and pasture farming* (Moscow-Leningrad, 1956)

Opportunities of short-term weather forecasts data in the environmental monitoring of the Kazakhstan

Alexey Terekhov [1,2*], *Nurlan* Abayev [2,3], *Kainar* Bolatov [2,3], *Zaure* Egemberdyeva[1,4],

[1] Institute of information and computing technologies MES, Almaty, 050010 Kazakhstan
[2] RSE "Kazhydromet", Almaty, 050022, Kazakhstan
[3] al-Farabi Kazakh national University, Almaty, 050040, Kazakhstan
[4] Almaty University of Power Engineering and Telecommunications, Almaty, 050013, Kazakhstan

Abstract. This paper considers the possibilities of assimilation of FEWS NET (Famine early Warning System Network) products in natural resource monitoring of the territory of Kazakhstan in the cold period. With an area of 2.7 million km^2, Kazakhstan is located in the center of Eurasia in the arid zone, and snow is one of the most important water sources in the first half of the vegetation season. The products "Snow Depth", "Temperature" and "Precipitation " developed for Central Asia are based on model meteorological data, including short-term weather forecasts. This information, presented in the form of regular matrices, has several advantages in efficiency and spatial detailing over direct ground (remote) measurements of meteorological parameters. The use of FEWS NET products for the environmental monitoring of the entire territory of Kazakhstan or its separate parts is of benefit for assessing the current situation on snow reserves, weather regimes and prospects for spring snowmelt.

1 Introduction

At present, the weather forecast is a highly demanded and developed area of information processing, in which huge estimated capacities are involved. The vast amounts of data from a developed network of ground-based meteorological stations and satellite-based sensing information are analyzed. To assimilate this information, a large number of models have been developed that describe and predict the main weather characteristics of the underlying surface of the Earth with different space-time scales [1,2]. As a result, the relevance of weather forecasts, especially short-term forecasts, is quite high. Predictive data of meteorological parameters have the format of regular matrices with a characteristic period of one kilometer and above. In the current information flow scheme, actual (ground-based and satellite) monitoring of the state of the underlying surface of the Earth is significantly inferior in regularity and information saturation to the model characteristics

* Corresponding author: aterekhov1@yandex.ru

based on short-term meteorological forecasts. This circumstance provides the basis for the creation of systems of natural resource territorial monitoring, based on short-term meteorological forecasting data, opening for them a new important and popular application.

The calculated model data of meteorological characteristics are available for free on various portals. The largest of them are global data assimilation systems [3,4], such as the GDAS (Global Data Assimilation System [https://www.ncdc.noaa.gov]), LDAS (Land Data Assimilation System). There are the regional products. Famine Early Warning System Network (FEWS NET) [https://earlywarning.usgs.gov/fews/overview] administers such activity for some regions in Africa, Asia, and Latin America.

The arid climate of Central Asia, including Kazakhstan, makes this territory very dependent on winter snow accumulation [5]. Snow is an important water resource, which provides the following: activity of spring floods [6]; humidification of arable land in the first half of the growing season and accordingly parameters affecting the state of vegetation cover and the yield of grain crops on the rain-fed agriculture [7-9]. The volume of runoff of large rivers of the region (River Ile, River Kara-Ertis) and the work of the hydraulic units located in transboundary river basins [10-11] is significantly affected by the snow accumulation in mountain conditions (the territory of Kazakhstan includes parts of Tien Shan, Jungarian Alatau and Altai). The purpose of this work is to demonstrate the capabilities of natural resource monitoring of the territory of Kazakhstan in the cold period, based on FEWS NET products.

2 Test site territory

Kazakhstan is the largest country in Central Asia. (an area of 2.72 million km^2, the 9th place in the world), fig. 1. The country is located in the center of Eurasia and has no access to an ocean. The climate is mainly arid, sharply continental. Almost half of the country's territory is semi-deserts and deserts climatic zones. The very low population density (6,8 people\km^2, 184th in the world) makes it very important to develop natural resource monitoring systems for the widest possible list of parameters.

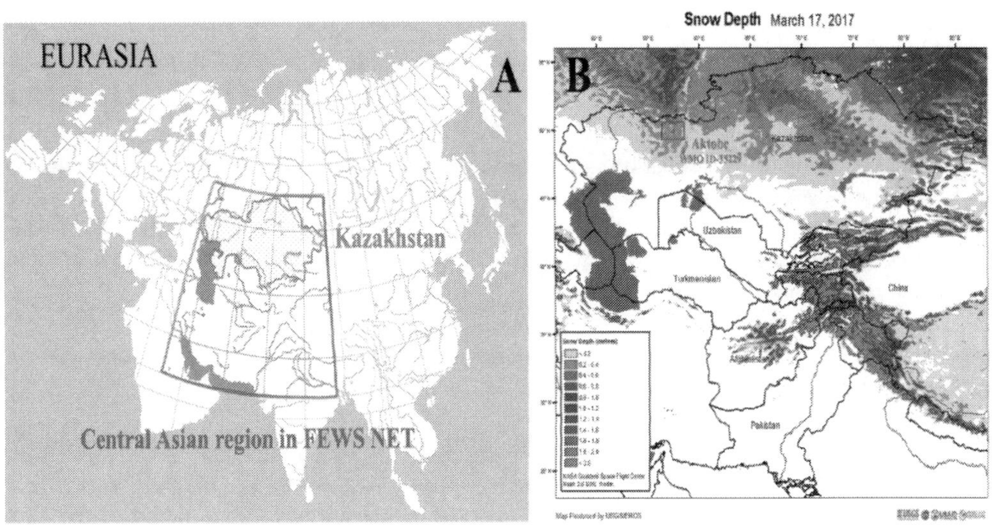

Fig. 1. A – Disposition of Central Asian region (FEWS NET) and Kazakhstan on Eurasia continent; B – example of daily "Snow Depth" map for Central Asia from FEWS NET with disposition meteorological station Aktobe, Kazakhstan.

3 Materials and methods

FEWS NET portal products presented for the Central Asia region (CA) were used as raw data. Three products are of the greatest practical interest in monitoring the territory of Kazakhstan during the cold period which are "Snow Depth (SD)", (fig.1b); "Temperature (T)"; "Precipitation (P)". These options are available as overview maps in geographic projection, (Lat/Lon) in "png" file format. For the SD product, the spatial resolution is 0.044° x 0.044°, with a daily update. Products " T "and" P" have a spatial resolution of 0.1423° x 0.1423°, with a ten days update. The depth of the SD archive linked to Terra MODIS satellite data is 20 years: From October 2000 to the present. The remaining products ("T" and "P") are available since 2013.

The ten-day product "Temperature" is an averaging of 6-hour forecasts of air temperature (at a height of 2 m). Map of deviations of the current temperature from the climate norm, which is constructed as a difference between the current values of ten-day temperature and the data of ground weather stations 1960-1990 is also interesting.

The ten-day value of the product "Precipitation" is calculated within the framework of the RFE (Rainfall Estimation) ver.2.0 from the NOAA Climate Prediction Center (CPC). The daily rainfall estimation is based on various satellite data and is adjusted according to the WMO ground network of weather stations. Daily estimates are aggregated to ten-day values.

4 Results

RSE "Kazhydromet" has been producing a ten-day pilot report titled "Snow coverage of Kazakhstan" since 2018. The report describes the weather conditions and characteristics of the snow coverage during the cold period, from November 1 to April 30. In the season of 2019-2020, the report is based on the FEWS NET products "Snow Depth", "Temperature", "Precipitation". The Figure 2c, 3 show the sample illustrations from report (No 3) of 30th of November, 2019.

5 Validation

In their development and application, short-term forecast meteorological products underwent numerous validation procedures that demonstrated their high accuracy. Validation of snow height for the territory of Central Asia is fragmentary. For the territory of Kazakhstan, there is information on validation for the zone of dry steppes and semi-deserts, which shows a good correspondence between "Snow Depth" and ground measurements [12], fig.2 (a,b).

6 Discussion

The overview presented in the report "Snow coverage of Kazakhstan" is of significant interest, fig.(2c,3). Since the data from meteorological stations, is relatively small, and because of the variability of climatic conditions of their locations (often foothills), the data cannot be formed in a comparable product in terms of efficiency and information provision. The existing data archive allows us to use not only absolute but also relative assessments, which increases the accuracy of estimations. The subsequent processing of FEWS NET

 https://doi.org/10.1051/e3sconf/202014903003

products for the purpose of monitoring the conditions of spring floods, which pose a real threat to the economic facilities and population of Kazakhstan, may become a promising direction.

7 Conclusion

Fields of the short-term forecast of meteorological parameters and data obtained using various meteorological models can be used as a basis for monitoring of natural resources of the territory of Kazakhstan, characterized by extremely low population density. The efficiency and information content of such products is much higher than the data of ground (remote) monitoring of environmental parameters. The use of relative estimates based on existing model information archives with a depth of up to 20 years reduces the importance of absolute errors and opens up new opportunities for monitoring practically important and/or dangerous phenomena, such as spring floods.

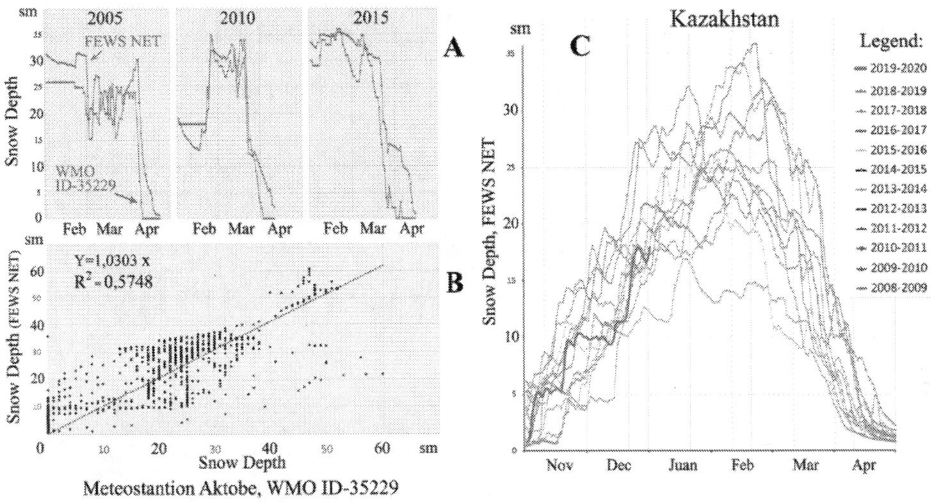

Fig. 2. A - Snow depth evolution during February – April for 2005, 2010 and 2015 years. Comparison meteorological station (Aktobe, WMO ID-35229) data with FEWS NET "SD" product.
B –Comparison snow depth daily data during February-April 2005-2019 years. Ground data (Aktobe station) vs FEWS NET "Snow Depth" product (averaging zone 150x130 km around meteorological station Aktobe. C - Seasonal dynamics of "Snow Depth" for territory of Kazakhstan create on base of daily "SD" data from FEWS NET.

 https://doi.org/10.1051/e3sconf/202014903003

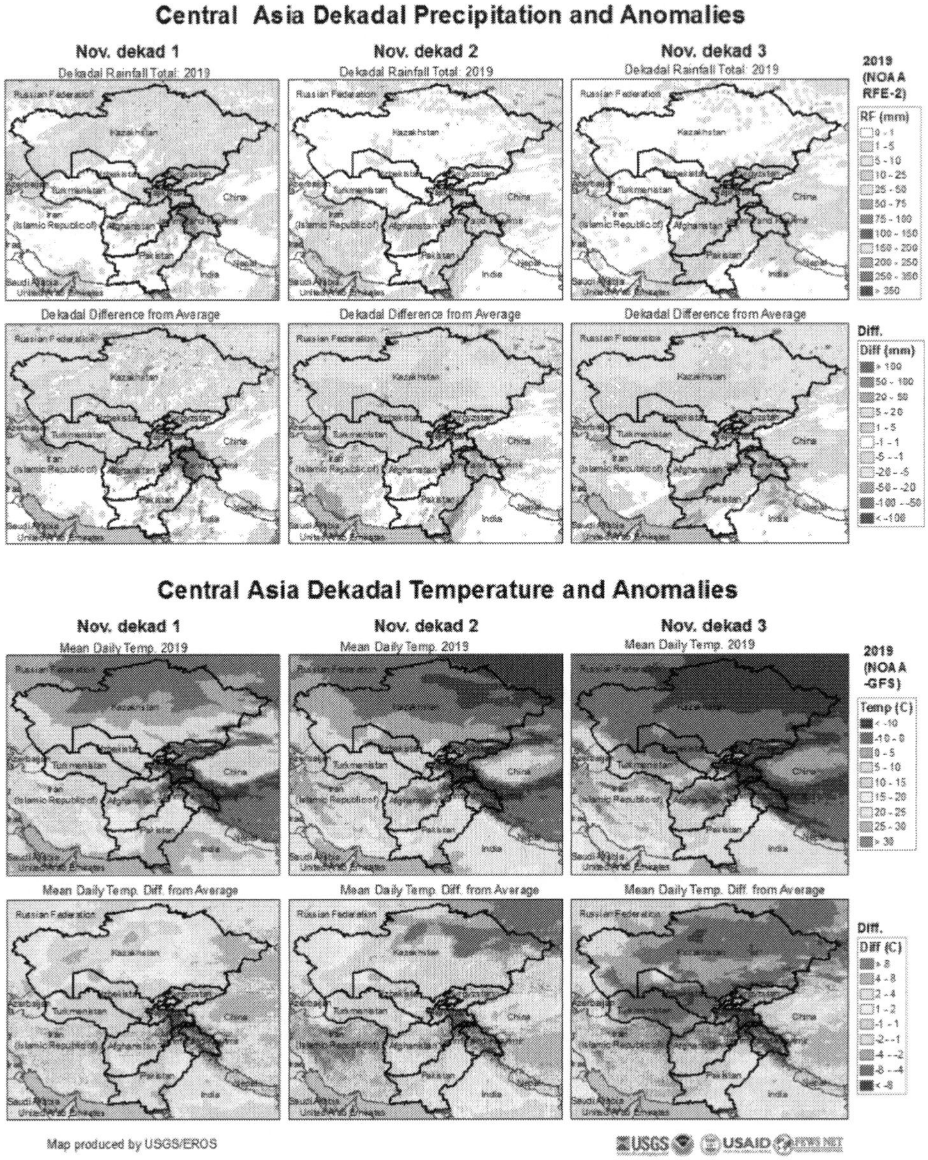

Fig. 3. Example of "Temperature" and "Precipitation" monitoring on base of decadal FEWS NET products (November, 2019).

Acknowledgment

This work was financially supported by Ministry of Education and Sciences of Kazakhstan, grant AP05134241.

References

1. T. N. Krishnamurti, C. M. Kishtawal,T. E. LaRow, D. R. Bachiochi, Z. Zhang, C. E. Williford, S. Gadgil, S. Surendran, Improved Weather and Seasonal Climate Forecasts from Multimodel Superensemble, *Science*, 1999, Vol. **285**, Issue 5433, pp.1548-1550, DOI: 10.1126/science.285.5433.1548.
2. T. N. Krishnamurti, C. M. Kishtawal, Z. Zhang, T. LaRow, D. Bachiochi, E. Williford, S. Gadgil, S. Surendran, Multimodel Ensemble Forecasts for Weather and Seasonal Climate, *J. Clim.*, 2000, Vol. **13**, No. 23, pp. 4196-4216, DOI: 10.1175/1520-0442(2000)013<4196.
3. L. Bengtsson, M. Ghil, E. Kallen, *Dynamic Meteorology: Data Assimilation Methods,* New York: Springer, 2012, 330 p, DOI: 10.1007/978-1-4612-5970-1.
4. W. Lahoz, B. Khttatov, R. Menard, *Data Assimilation: Making Sense of Observations,* Heildelberg: Springer; 2010, 718 p, DOI: 10.1007/978-3-540-74703-1.
5. A. G. Terekhov, N. N. Abayev, N. R. Yunicheva, Anomalous snowy regime at 2019 year and long-term trends in snow depth in Kazakhstan, *Sovremennye problemy distantsionnogo zondirovaniya Zemli iz kosmosa*, 2019, Vol. **16**, No. 5, pp. 351–355, DOI: 10.21046/2070-7401-2019-16-5-351-355.
6. A. G. Terekhov, I. T. Pak, S. A. Dolgikh, Satellite observations of the anomalous spring flood at the lower reach of the Ayaguz river in 2016, *Sovremennye problemy distantsionnogo zondirovaniya Zemli iz kosmosa*, 2016, Vol. **13**, No. 4, pp. 273–276, DOI: 10.21046/2070-7401-2016-13-4-273-276.
7. U. Sultangazin, N. Muratova, P. Doraiswamy, A. Terekhov, Estimation of weed infestation in spring crops using MODIS data, *Proceedings of IGARSS'03*, Toulouse, 2003, Vol. **1**, pp. 392-394.
8. N. Muratova, A. Terekhov, Estimation of spring crops sowing calendar dates using MODIS in Northern Kazakhstan, *Proceedings of IGARSS'04*, Anchorage, 2004, Vol.**6**, pp. 4019-4020.
9. L. Spivak, I. Vitkovskaya, M. Batyrbayeva, A. Terekhov, The experience of land cover change detection by satellite data, *Frontiers of Earth Science*, 2012, Vol. **6**, No. 2, pp. 140-146. DOI: 10.1007/s11707-012-0317-z.
10. A. G. Terekhov, I. T. Pak, S. A. Dolgikh, Hydrology monitoring of Kapchagay reservoir on river Tekes (China's part of river Ile basin) based on LANDSAT 5,7,8 data and DEM batymetry, *Sovremennye problemy distantsionnogo zondirovaniya Zemli iz kosmosa*, 2015, Vol. **12**, No. 6, pp. 174–182.
11. A. G. Terekhov, A. A. Pak, Influence of the Kapshagay reservoir (China) refill on transboundary River Ile runoff and satellite-based forecasting, *Sovremennye problemy distantsionnogo zondirovaniya Zemli iz kosmosa*, 2019, Vol. **16**, No. 4, pp. 298–302, DOI: 10.21046/2070-7401-2019-16-4-298-302.
12. A. G. Terekhov, N. I. Ivkina, N. R. Yunicheva, I. S. Vitkovskaya, A. G. Yeltay, Changes in seasonal snow cover in dry steppes and semi-deserts of Kazakhstan: an example using data between 2001 and 2019 from Emba River basin, *Sovremennye problemy distantsionnogo zondirovaniya Zemli iz kosmosa*, 2020, Vol.**17**, in press.

The effect of changing stratification in the atmosphere in central zone of Eurasia according to vegetation data of Tien Shan mountains during 2002-2019

Alexey Terekhov[1,2], *Nurlan* Abayev[2,3], and *Irina* Vitkovskaya[1]

[1] Institute of information and computing technology MES, Almaty, 050010, Kazakhstan
[2] RSE Kazhydromet, Almaty, 050022, Kazakhstan
[3] al-Farabi Kazakh National university, Almaty, 050040, Kazakhstan

Abstract. On the basis of long-term dynamics of the July's vegetation data (NDVI and VCI indices) for 31 ranges of the Tien Shan and Jungarian Alatau arid mountains located in the Eurasia center are found the atmosphere stratification change. The variability of Atlantic Ocean water vapor transport ability over Northern Tien Shan ranges to the Inner Tien Shan was detected. This phenomenon not related to the overall seasonal humidity but is obviously associated with atmosphere stratification regimes. The analyzed period consists of two eras from 2002-2008 and 2008-2019 years, which differ in the regime of accessibility of the Inner Tien Shan ranges for water vapor from Atlantic Ocean. The overall tendency of the changes in 2002-2019 is the increase in the availability ocean water vapor to the Inner Tien Shan ranges. Recorded changes may be due to global air temperature increases and atmospheric processes intensification.

1 Introduction

Atmospheric patterns transporting moisture from the Atlantic Ocean to the inlands of Eurasia are in complex interaction [1] with arid mountain ranges located in the center of the continent. Part of the ranges situated in the front edge of the dominating atmospheric transportation, capture the most of the moisture. Inner ranges, as a rule, are characterized by a significantly more arid climate [2]. Changes in atmospheric stratification affect the moisture conditions of the inner ranges of mountains land, which are expressed both in changes in snow deposite [3], and in vegetation condition [4,5]. As for vegetation cover, it is worthwhile noting the presence of two factors. The first one is in charge of absolute changes in the vegetation state. This is mainly due to the amount of water vapor transported and, partly, to temperature regimes. The second factor is traced by the relationship between vegetation state of the front and inner ranges. This ratio is related to the changes in the atmospheric stratification condition.

The ability of atmospheric flows to transport water vapor through mountain ranges cropping up along the way, in general, depends on width of the area [2]. This ability is stronger in the area of ascending branches of atmospheric circulation cells (Hadley and Ferrel cells), and weaker in the area of descending branches, respectively [6]. Climate changes, especially intra-continental ones, are almost completely controlled by atmospheric condition [7]. Atmospheric stratification regimes and its temporal variations are one of the most crucial parameters in the Earth's climate system. The ability to obtain information on variability of atmospheric stratification in the central zone of Eurasia from satellite monitoring of vegetation state of separate mountain ranges [4] and intermountain valleys [5] seems to be very relevant for understanding the tendency of the climate changes.

The state of natural vegetation cover is one of the indicators illustrating temperature and humidity conditions of underlying surface of the Earth. The central zone of Eurasia is distinguished by arid climate where the key parameter determining the vegetation state, is moisture conditions. Satellite observations using specific vegetation indices, of which NDVI [8] being the most popular, enable to obtain consistent sets of observations and use them to further determine current trends in changes of weather conditions at various spatial and temporal scales. Analysis of mountain range vegetation allows diagnose temperature and humidity conditions of underlying surface and their dependence on height above sea level and location inside the mountain lands [4].

The purpose of this study is to diagnose the changes in atmospheric stratification condition in the central zone of Eurasia in 2002-2019 based on data on vegetation state in 31 ranges of Tien-Shan and Dzungarian Alatau, located within 40^0-46^0 N latitude, 70^0-90^0 E longitude.

2 Research territory

The mountain ranges of Tien-Shan and Dzungarian Alatau adjoining to it on the north, are situated in the center of the Eurasian continent and occupy a territory of approximately 2500x600 km, Fig.1(a). The highest point of Tien-Shan is Pik Pobedy - 7439 m above sea level. Tien-Shan land consists of distinct ranges with characteristic heights of dominant peaks between 4000-6000 m. Geographers attribute the ranges to various groups, Fig.1(b). The main of which are Northern, Eastern, Western and the Inner (Central) Tien-Shan. The mountain ranges of the Northern Tien-Shan (Kyrgyz (No.6), Trans-Ili Alatau (No.7), Kungey Alatau (No.8) and Ketmen (No.9) (Fig.1(b)) are at the front (north-west) edge of this mountains and act as screens for the rest of the ranges, in particular, the Inner Tien-Shan, in relation to air masses, transporting water vapor from the Atlantic Ocean. The ranges of Tien-Shan significantly dry the overpassing atmospheric flows; therefore, one of the largest sand deserts of the world, Takla-Makan (1000x400 km), adjoins the Tien Shan on the south, with annual precipitation of less than 50 mm.

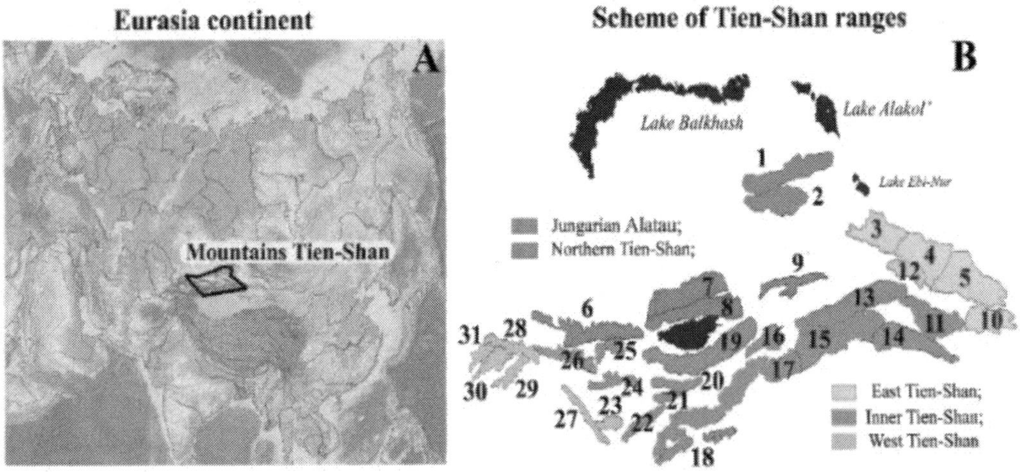

Fig.1. A - Disposition of mountains Tien-Shan in Eurasia continent; B- scheme of mountain ranges (1-31) used as objects for monitoring of the vegetation state.

3 Materials and methods

3.1 The satellite data

Temporally Smoothed NDVI, e-MODIS NDVI C6 product has been used in the study, which is publicly available on the website of Early Warning and Environmental Monitoring Program [http://earlywarning.usgs.gov] [9]. The U.S. Geological Survey (USGS) Earth Resources Observation and Science (EROS) Center distributes satellite-derived vegetation products generated from the Collection 6 Moderate Resolution Imaging Spectroradiometer (MODIS) instrument flown aboard the Aqua satellite. These products respond to operational land monitoring applications requiring near-real time Normalized Difference Vegetation Index (NDVI) data for comparison against historical records. Real-time and historical NDVI products are composited in 10-day (decadal) intervals on a Geographic-mapping grid (Spheroid: WGS84). Archive depth is 18 years, from 2002 till present. e-MODIS NDVI C6 scene for Central Asia has the following dimensions (17407x13676 pix), with pixel size: x-direction: 0,002 (deg); y-direction: 0,002 (deg). The scene covers the territory within approximately 24-56°N latitude, 47-87° E longitude, including the Tien-Shan mountain system. The product features special spatial and temporal filters [9], which enables to minimize interfering effect of cloudiness. The seasonal characteristic of mountain vegetation was based on two scenes, decade 2 and 3 (decade 20 and 21). This period refers to the season maximum of mountain ridge vegetation observed in the second half of July.

3.2 Methods of satellite data processing

Intermountain plains separating of Tien-Shan lands are characterized by a semi-desert climate with low precipitation. Therefore, natural vegetation of Tien-Shan mountains represents an "island" of vegetation amid the mountain semi-deserts. Similar islands of vegetation, which are shaped due to the increased precipitation on mountain slopes, are easily mapped. Vegetation masks in 31 ranges of Tien-Shan and Dzungarian Alatau have been created using expert decoding, Fig.1(B). Vegetation masks include subalpine/alpine meadows and mountain forests. Description of the seasonal vegetation state was based on two scenes, decade 20 and 21 (July 10-31), from which a matrix of accumulated maximum values at each pixel position has been produced. As a characteristic of the year of vegetation masks of each of the analyzed ranges, the average NDVI value for the contour has been used.

The vegetation cover of separate ranges Tien-Shan exhibits various numerical NDVI values and their multi-year variability. To bring all analyzed objects to comparable scale, NDVI values have been restated into the values of VCI (Vegetation Condition Index) [10,11]. In this respect, the modified version thereof has been applied [12], with three reference fixed values, where: 0 – multi-year low NDVI; 0,5 – arithmetic mean of a number of NDVI (2002-2019) samples; 1,0 – multi-year high NDVI. Additional fixation of the average value makes the comparisons of the conditions of separate ranges more accurate from the meteorological perspective, where the average value (norm) is one of the key concepts. For ease of presentation, VCI values, as a rule, are multiplied by 100, which forms a scale in the range of 0-100.

Transfer to the comparable vegetation state assessment scale enabled to diagnose the relevant variability of vegetation condition on different Tien-Shan ranges. For this purpose, for each season, a difference between [VCI_{range}] and [$VCI_{total\ mask}$] has been calculated. Time series (2002-2019) of these differences contain information about relevant changes in vegetation condition within different parts of the mountain lands, which, in turn, depends on the atmospheric stratification condition of this region.

4 Results

Changes in vegetation condition of the analyzed mountain ranges, in general, are represented in Fig.2. The conditions of relevant changes in vegetation of separate ranges, calculated across time series [VCI_{range}]-[$VCI_{total\ mask}$] during the period of 2002-2019, are shown in Fig.3. To illustrate the trends of the relevant vegetation changes of the analyzed ranges, linear trend parameters have been used, in particular, the sign and value of the coefficient in the linear trend equation and the approximation confidence value (R^2).

Such approach enables to differentiate between the dynamics of differences in vegetation conditions of separate ranges of Tien-Shan and the regional vegetation state. The calculation results are shown for three time periods: 2002-2019; 2002-2008; 2009-2019. The legend on the presented map diagrams reflects the relationship between the values of the main parameters of the linear trend of [VCI_{range}]-[$VCI_{total\ mask}$] time series. The legend is designed in such a way that more intense colors correspond to high R^2 values. At the same time, red color is attributed to those objects where there is a negative trend in the time series of the calculated VCI value, blue - to objects with a positive trend.

The relationship between the relative vegetation condition in Zailiisky Alatau and the general vegetation condition of the entire analyzed mountain range has also been assessed, Fig. 4.

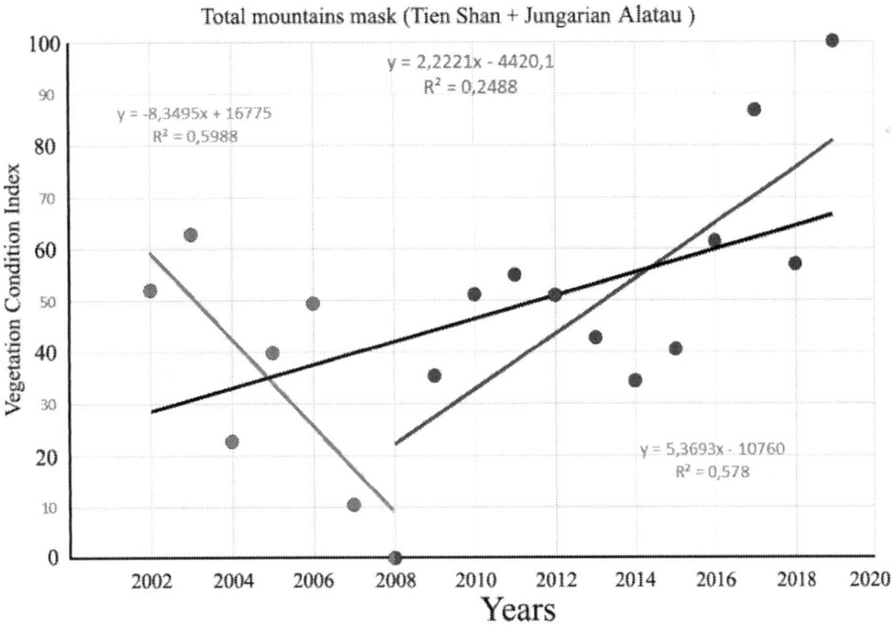

Fig. 2. Dynamics of VCI values for total mask which included all ranges (Tien Shan and Jungarian Alatau) during 2002-2019. Two time periods are highlighted in color: 2002-2008 (red); 2008-2019 (blue). The values of parameters of linear trend equations for the periods: 2002-2008 (red); 2008-2019 (blue); 2002-2019 (black); are given.

5 Discussion

Atmospheric flows transporting water vapor from the Atlantic Ocean to the center of the continent are under the influence of various factors, and their seasonal characteristics are quite variable. This leads to year-to-year changes in temperature and moisture conditions that reflects in vegetation state of Tien-Shan and Dzungarian Alatau ranges. The main weather factor affecting the mountain vegetation state in July is moisture (air humidity, amount of precipitations). Humid years alternate with dry years, but at the same time, noticeable persistence is recorded in multi-year perspective. During the period of 2002-2008 there was recorded a deterioration in vegetation condition, Fig.2. Then, from 2008 to 2019, vegetation state has improved, Fig.2. Relative vegetation condition of separate ranges during the period of 2002-2008 had a clear structure, Fig.3. The front edge of Tien-Shan, attributed to the Northern Tien-Shan (ranges No. 6-9), where there is more intense precipitation, showed much stronger deterioration than other ranges. For the period of 2008-2019, mild stratification of the territory is characteristic, with the exception of two ranges of the Inner Tien-Shan: Kungey Alatau (No. 19) and the Borokhotan ridge (No. 13), which have demonstrated marked

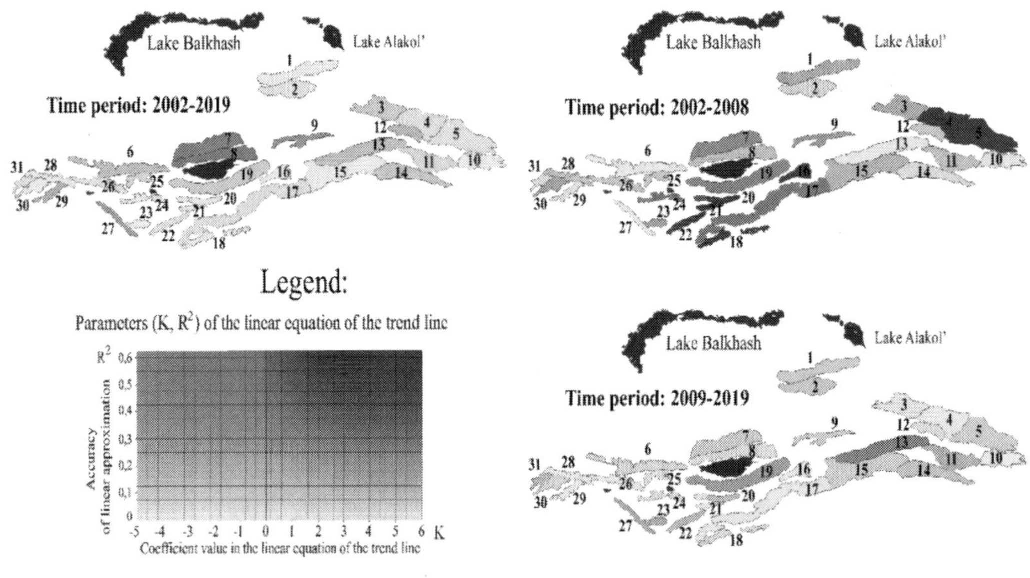

Fig. 3. The schemes of Tien-Shan and Jungarian Alatau mountain ranges, which colored by linear trend parameters (the coefficient [K] of the trend equation and the confidence value of the linear approximation [R^2]) are presented. The time series ($VCI_{range} - VCI_{total\_mask}$) during 2002-2019 served as the basis for this analytical procedure. The characteristics of three time periods: 2002-2019; 2002-2008; 2009-2019; are shown.

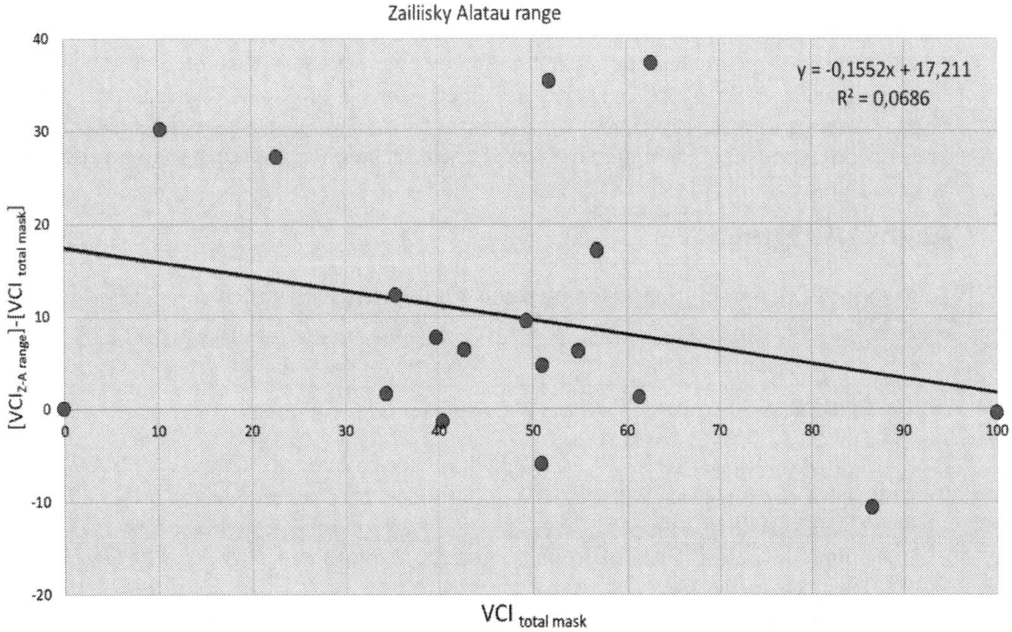

Fig. 4. The relation the July's vegetation state of all mountain ranges (VCI $_{\text{total mask}}$) against the differences between July's vegetation state of the Zailiisky Alatau range (VCI $_{\text{Z-A range}}$) and all ranges (VCI $_{\text{total mask}}$) during 2002-2019 years.

deterioration in vegetation condition, Fig.3. the entire monitoring period of 2002-2019 in total had a tendency to relative deterioration in vegetation condition of the Northern Tien-Shan ranges, Fig. 3. The foundation for this effect has been laid in 2002-2008.

The results obtained show that of the Northern Tien-Shan ranges are distinguished into a separate group showing relative deterioration in vegetation state. Whereby no clear relationship between the degree of relative deterioration in vegetation condition of the Northern Tien-Shan ranges, for instance, Zailiiskiy Alatau (No. 7) and vegetation conditions of the entire analyzed region, Fig.4, has been observed. I.e. the ability of atmospheric patterns to transport water vapor to the Inner Tien-Shan is not related with the overall seasonal moisture.

The ability of atmospheric patterns to transport water vapor through of the Northern Tien-Shan ranges is determined by atmospheric stratification conditions. The reasons of multi-year changes in the characteristics of these conditions remain unknown. It is possible that this phenomenon is related to the changes in the width of global atmospheric circulation cells (Ferrel, Hadley) because of its reaction to changes in the energy balance of the Earth.

6 Conclusion

The analysis of NDVI and VCI vegetation indices in 31 ranges of Tien-Shan and Dzungarian Alatau during the period of 2002-2019 has shown that there exists change in the ability of atmospheric flows to transport water vapor from the Atlantic Ocean towards the center of Eurasia to the ranges of the Inner Tien-Shan. This ability varies at various monitoring periods. Time span of 2002-2019 can be conditionally represented as consisting of two stages. The first, the period of 2002-2008, which is characterized by a marked increase in the availability of water vapor transported towards the ranges of the Inner Tien-Shan. The second, the period of 2008-2019, with a stable pattern that does not have clear trends. The general tendency of the analyzed period shows an increase in the share of water vapor transported by atmospheric processes to the Inner Tien-Shan. The recorded effect is obviously associated with the changes in atmospheric stratification in the center zone of Eurasia, which, in turn, is a consequence of the reaction of global atmospheric circulation to the Earth's climate changes.

Acknowledgment

This work was financially supported by Ministry of Education and Sciences of Kazakhstan, grant AP05134241.

References

1. E. M. Aizen, V. B. Aizen, J. M. Melack, T. Nakamura, T. Ohta, Precipitation and Atmospheric circulation patterns at mid-latitudes of Asia, *Int.J.Climatol.*, 2001, **21**, pp.535-556, DOI: 10.1002/joc.
2. R. B. Smith, The Influence of Mountain on the Atmosphere, *Advances in Geophysics,* 1979, Vol.**21**, pp.87-230.
3. A. G. Terekhov, A. A. Pak, Influence of the Kapshagay reservoir (China) refill on transboundary River Ile runoff and satellite-based forecasting, *Sovremennye problemy distantsionnogo zondirovaniya Zemli iz kosmosa*, 2019, Vol. **16**, No. 4, pp. 298–302, DOI: 10.21046/2070-7401-2019-16-4-298-302.
4. A. G. Terekhov, I. S. Vitkovskaya, N. N. Abayev, S. A. Dolgikh, Long term trends in vegetation in Tien-Shan and Dzungarian Alatau from eMODIS NDVI C6 data (2002–2019), *Sovremennye problemy distantsionnogo zondirovaniya Zemli iz kosmosa*, 2019, Vol. **16**, No. 6, pp. 133–142, DOI: 10.21046/2070-7401-2019-16-6-133-142.
5. A. G. Terekhov, N. N. Abayev, E. I. Lagutin. Diagnostic of water availability of agricultural culture in Xinjiang during 2003-2019 years on base of eMODIS NDVI C6, *Sovremennye problemy distantsionnogo zondirovaniya Zemli iz kosmosa*, 2020, Vol.**17**, in press.
6. E. N. Lorenz, *The Nature and Theory of the General Circulation of the Atmosphere.* World Meteorological Organization, 1967, 109 p.
7. J. Huang, M. B. Mc Elroy, Contributions of the Hadley and Ferrel Circulations to the Energetics of the Atmosphere over Past 32 Years, *J. Clim.* 2014, Vol.**27**, pp. 2656-2666, DOI: 10.1175/JCLI-D-13-00538.1.
8. J. Rouse, R. Haas, J. Schell, Deering D. Monitoring vegetation systems in the great plains with ERTS. *Third ERTS Symp. NASA SP-351,* 1973, Vol. **1**, pp. 309-317.
9. D. Swets, C. Bradley, J. Rowland, S. E. Marko, *A weighted least-squares approach to temporal NDVI smoothing*, 1999, URL: https://pubs.er.usgs.gov/publication/70201050.
10. F. N. Kogan, Application of vegetation index and brightness temperature for drought detection, *Advances in Space Research,* 1995, Vol.**15**, Issue 11, pp. 91-100, DOI: 10.1016/0273-1177(95)00079-T.
11. F. Kogan, Remote Sensing of Weather Impacts on Vegetation in Non-Homogeneous Areas, *Int. J. Remote Sens.*, 1990, Vol.**11**, pp. 1405–1419.
12. A. Terekhov,N. Makarenko, I. Pak, Composite index of temperature conditions and its applicability to the description of the state of agricultural vegetation in Kazakhstan, *Sovremennye problemy distantsionnogo zondirovaniya Zemli iz kosmosa*, 2015, Vol. **1**, pp. 31-40.

https://doi.org/10.1051/e3sconf/202014903005

Estimation of the vegetation diversity of the Western Tannu-Ola Range

Khulermaa B. Kuular[1]
Tuvinian Institute for Exploration of Natural Resources SB RAS, ul. Internatsionalnaya 117a, Kyzyl, 667007, Russia

Abstract. This paper discusses the main indicators diversity of vegetation in key areas of the northern macro slope of the Western Tannu-Ola Range. The degree of forest cover and vegetation fragmentation were estimated using Landsat-TM/ETM+/OLI images for June 25, 1992, June 18, 1998 and June 24, 2015 in QGIS using the plugin Landscape Ecology. The Simpson index shows of increasing biodiversity in different altitudes. The results showed that with increasing climate warming in the region and maintaining this trend, the proportion of dark coniferous forest at all altitude levels (foothill, mid-elevation and high-elevation zones) will increase. And this led to a decrease in the fragmentation of the forest cover of northern macro slope of this range.

1 Introduction

Within the framework of "The Biodiversity Conservation" were recommended to assess ecosystem diversity. For assess ecosystem diversity it was proposed to use the ratio of the areas of different forest types relative to the total area, as well as the fragmentation of forest cover [1]. Among the priority tasks is the assessment of mountain biodiversity.

Currently, the impacts to climate change in territory of Central Asia are manifested noticeably. The mountainous ecosystems of Southern Siberia are remarkably sensitive to the vulnerable environment conditions. The most serious among them are the processes of change in the southern boreal forests. Central Asia is characterized a variety of plant and climatic conditions. Taking into account and assessing the diversity of forest communities in the republic remains a problem due to the limited availability of a significant part of the region. It is possible to evaluate the values indicators the diversity of vegetation in hard-to-reach mountain areas which employed remote sensing techniques.

The purpose of this study is assessing the forest cover and fragmentation on key areas of the northern macro slope of the Western Tannu-Ola Range using data of the Landsat-TM/ETM+ and Landsat- OLI.

2 Material and Method

The study areas are situated in the territory of the Western Tannu-Ola Range within the boundaries of 50 ° 55'-51° 15'N and 91°30'-92°20'E. The Western Tannu-Ola range occupies in the southwestern part of the republic and plays an important role in ensuring the sustainability of the mountains forests of Southern Siberia. The range is one of the large climatic borders, the final barrier on the way of moist northwestern airflows which have reached the mountains of South Siberia.

The study areas are characterized by sharply continental climate, with annual average precipitation of 229.8±65.1 mm and annual mean temperature of is -1.0±0.8°C (1988-2018). An anomaly of the mean annual temperature amounted to 1.4±0.1°C compared with the base period (1961-1990) [2].

The Western Tannu-Ola Range own highly heterogeneous characteristics with a large variety of climate, vegetation and soil. It is characterized by a unique combination of different landscapes – from dry mountain steppes to tundra [3]. This diversity of vegetation helps maintain the stability of the range vegetation. The main

[1] Corresponding author: k-k-188@list.ru

forest-forming species are larch (*Larix sibirica*), cedar (*Pinus sibirica*), spruce (*Picea obovata*), and birch (*Betula microphylla and Betula rotundifolia*).

The processing was performed in the Quantum GIS of data Landsat-TM/ETM+/OLI. A band combination 5, 4 and 3 Landsat-TM/ETM+ and 6, 5 and 4 Landsat-OLI were used for image analysis, where different classes have been identified. Information for identification was obtained from the forest scheme-map of the Tyva Republic (scale 1: 300 000, 1992).

Three altitude levels (foothill, mountain-taiga and high-mountain zones) were identified in the study areas. The foothill zone (1100-1300 m above sea level) - the slopes bordering the steppe are covered with larch forests with birch. In the mountain-taiga (1300-1700 m above sea level) are dominated by the Siberian cedar and Siberian larch. Cedar (5C3L2S) is dominates in the composition of the 1-st tier, where the conditions are favorable for it. In drier areas larch is dominates (7L3C+S). Dark coniferous species (4C4S1L, 5C5S) are dominated in the composition of 2d tiers. In the high-mountain zones (1700-2120 m above sea level) larch and cedar forests grow. Here mainly grows dark coniferous forest and cedar is dominates in the composition of 1 and 2 tiers (10C, 7C3L).

Within the key areas the vegetation was divided into the main classes of the underlying surface: 1 - dark coniferous forests (cedar), 2 - light coniferous forests (larch), 3 - dark coniferous forests (spruce), 4 - small-leaved leaves (birch), 5 - non-forest areas, 6 - natural grassland, 7 - wetlands/tundra, 8 – sparse growth, 9 - shrubs, 10 - water bodies (rivers, lakes), 11 - new forest areas / reafforestation.

The identified classes is verified in cross examination way at present context, whether the category exists or not or converted into another pattern. Diversity of vegetation was calculated based on the above vegetation classification using the plugin Zdetsaka. The diversity indices (forest cover, the number of fragments, and the Simpson index)) were estimated fusing the plugin Landscape Ecology.

3 Result and Analysis

Forest integrity which is defined as the ratio of the area covered by forest vegetation to the total area of forest land, as well as the proportion of dark coniferous forests, the proportion of light coniferous forests, and the proportion of small-leaved forests characterize the ecosystem diversity of study areas forests. Forest cover is described by the following formula:

$$EP = \frac{Sa}{S} \cdot 100 \quad (1)$$

where S – total area of forest land, Sa – area of forest.

An assessment of forest cover based on the percentage of forest covered area of key plots on each elevation levels (table 1). The table shows that the shares of dark coniferous forests with a predominance of spruce (class 3), small-leaved forests (class 4) and shrubs (class 9) increased at all altitudes levels compared to previous image data.

Table1. Forest cover (%) of classes of the Western Tannu-Ola Range

№ class	Foothill zone			Mountain-taiga zone			High-mountain zone		
	I	II	III	I	II	III	I	II	III
1	2.8	2.8	1.0	4.2	5.3	1.8	8.6	14.6	3.5
2	7.2	4.9	8.6	6.3	2.4	3.2	6.5	1.2	1.9
3	23.3	25.1	28.6	32.7	32.9	49.0	12.7	12.8	28.0
4	24.3	23.6	26.6	14.4	19.1	19.9	20.4	23.6	27.0
5	12.7	18.5	14.8	4.3	7.1	5.5	10.2	16.0	10.0
6	1.6	0.5	2.7	1.1	0.6	1.3	1.4	0.5	2.1
7	7.5	5.1	3.6	5.8	7.5	5.1	11.5	10.7	7.0
8	2.7	3.4	5.0	2.3	1.4	5.5	6.8	6.5	8.1
9	3.0	3.6	8.1	2.0	2.7	7.8	4.6	6.0	11.7
10	0.8	0.4	0.5	0.9	0.3	0.4	0.8	0.1	0.3
11	14.1	12.0	0.4	26.1	20.6	0.6	16.3	7.9	0.3

Designations: I – за дату 25.06.1992, II - 18.06.1998, III - 24.06.2015

Fragmentation of landscapes is an essential factor in reducing biodiversity [4; 5; 6]. Fragmentation of the forest cover is the division of entire forests into isolated fragments. From Table 2, it is seen that summarizes fragmentation of each elevation levels. The total fragmentation at three elevation levels was decreased on average 3 times more, of coniferous forests - 2 times more, of light coniferous forests - 4 times more.

Fragmentation of small-leaved forests occurs unevenly, but compared to first image, fragmentation also was decreased 1.3 times more. In general, the decrease of forest fragmentation is associated with overgrowth of the range forests with the establishment of favorable conditions during climate warming.

Table 2. Fragments on the key areas of the Western Tannu Ola Range

№ class	Foothill zone			Mountain-taiga zone			High-mountain zone		
	I	II	III	I	II	III	I	II	III
1	3274	3055	1308	3274	3441	1813	3274	3441	1813
2	5155	3827	1939	5155	2037	1093	5155	2037	1093
3	2922	3014	2024	2922	3545	1190	2922	3545	1190
4	3864	5631	2575	3864	6860	3202	3864	6860	3202
5	1120	1689	669	1120	2256	720	1120	2256	720
6	635	384	1883	635	404	1126	635	404	1126
7	987	1266	1969	987	975	1619	987	975	1619
8	2045	1650	2254	2045	1712	1871	2045	1712	1871
9	2188	2303	2750	2188	1692	2598	2188	1692	2598
10	119	227	95	119	201	83	119	201	83
11	2735	4198	154	2735	3528	137	2735	3528	137

Designations: I - 25.06.1992, II - 18.06.1998, III - 24.06.2015

The Simpson index had been a similar trend. The values of Simpson index for each altitude level are presented in the table 3. The Diversity Simpson Index [7], which takes into account both wealth and uniformity when measuring diversity. The Simpson diversity index is described by the following formula:

$$C = \sum \frac{n_i}{N} \tag{2}$$

where ni – the significance rating of each species (abundance or biomass), N - the sum of the significance ratings.

From Table 3, it is seen that that the Simpson index decreases which shows an increase in the diversity of vegetation in the key areas of the West Tannu-Ola range. It was shown that over the study period the fragmentation of the studied areas decreased which is associated with forest overgrowth during climate warming.

Table 3. The Simpson index of the Western Tannu-Ola Range

Simpson index	25.06.1992	18.06.1998	24.06.2015
Foothill zone	0.84	0.82	0.81
Mountain-taiga zone	0.87	0.86	0.81
High-mountain zone	0.87	0.67	0.81

From Table 4, it is seen that the change in the area of the studied key areas. In foothill zone the area of small-leaved forests with a predominance of birch increased 1.1 times more, in the mountain-taiga - 1.4 times more, in the high-mountain zones - 1.3 times more. In the foothill zone the area of dark coniferous forests with the predominance of spruce increased 1.2 times more, in the mountain-taiga - 1.5 times more, in the high-mountain zones - 2.2 times more; may be, part area of cedar forest is included here. In foothill zone the area of light coniferous forests increased 1.2 times more. In the mountain-taiga and high-mountain zones the light coniferous forests decreased (1.4 - 4.6%) probably because of the wildfire.

Table 4. Forest area (ha) of key areas of the Western Tannu Ola Range

№ class	Foothill zone			Mountain-taiga			High-mountain zones		
	I	II	III	I	II	III	I	II	III
1	5664	5736	1977	7783	10012	3291	19191	32371	7730
2	14619	9999	17402	11809	4538	6006	14513	2655	4297
3	47120	50684	57837	61313	61706	91767	28302	28437	62188
4	49099	47709	53793	26945	35852	37295	45231	52359	59972
5	25603	37451	29949	8055	13358	10337	22574	35627	22139
6	3297	1088	5531	1968	1107	2473	3216	1100	4626

7	15083	10399	7318	10839	13967	9465	25591	23791	15602
8	5367	6837	10150	4287	2562	10215	15159	14468	18072
9	6044	7242	16327	3712	5135	14639	10187	13433	26037
10	1695	758	1010	1710	520	742	1809	225	698
11	28528	24215	824	48958	38624	1149	36262	17569	674

Designations: I – за 25.06.1992, II - 18.06.1998, III - 24.06.2015

4 Conclusion

The calculation of landscape indices has been (the proportion of forests, the number of fragmentations and the Simpson index) in key areas of Western Tannu-Ola to assess changes in southern mountain forests for 1992, 1998 and 2015 using remote sensing data. An analysis of this data showed that over the past 23 years there has been a decrease in forest fragmentation and has led to an increase all elevation levels of forest on the Western Tannu-Ola ridge. The results showed a positive relationship between the growths of vegetation diversity in response to climate warming.

References

1. A.S. Isaev. *Monitoring the biological diversity of Russian forests: methodology and methods*. (2008).
2. V.N. Smagina, S.A. Ilyinskaya, D.I. Nazimova, I.F. Novoseltseva, Yu.S. Cherednikova. *Types of mountains forests of the Southern Siberia*. (Sciences, Novosibirsk, 1980).
3. Kh.B. Kuular. Sovr. Probl. DZZ Kosm. **15(7)**. 67–77. (2018).
4. L. Fahrig. Ann. Rev.ecol., evol.&syst. **34**. 487–515. (2003).
5. J. Fischer, D.B. Lindenmayer. Global Ecol. Biogeogr. **16(3)**. 265–280. (2007).
6. J. Wu. Princ. Guide Ecol. 438–444. (2009).
7. E.H. Simpson. Nature. **163**. 688. (1949).

Geoecological assessment of the state of postagrogenic lands of the Moscow village soviet council of the Republic of Khakassia on the basis of remote sensing data

Ekaterina V. Pavlova[1], *Anastasiia* I. Volkova[1], and *Ekaterina* A. Demina[1]

[1] Institute of natural sciences and mathematics, Khakass state University. N. F. Katanov 655000, Russia Federation, Abakan, Leninave, 90

Abstract. Currently, the consequences which take place in Khakassia expansion of tree-shrub vegetation on fallow lands have not been properly assessed neither from an ecological nor economic point of view. Based on the analysis of the agricultural map scale 1: 100 000 decoding images Landsat 4-5, 7, 8 and Sentinel 1, and 2, as well as subsatellite ground researches were carried out the identification, the description and assessment of the qualitative state of postagrogenic lands of Khakassia exposed to the processes of overgrowth of tree-shrub vegetation. As an example, this article analyzes the processes of overgrowth of agricultural land on the example of the territory of the Moscow village council of Ust-Abakan district. A geoinformation project of spatial distribution of postagrogenic lands within the Moscow village council of Ust-Abakan district of Khakassia was developed. The results of the research showed that in the studied area in the structure of agricultural land 67204 hectares of land belongs to the fallows located at different stages of recovery of which 77 % exposed to overgrowth processes. The obtained data indicate the need for the formation of management decisions in the field of land use.

Formation and maintenance of optimal land use structure is an urgent problem for many regions of Russia today. The processes of land privatization, which occurred in the 90s of XX century, led to the emergence of «ownerless» land plots that are not used for agriculture. Quite often, the owners of agricultural land do not comply with the requirements of legislation on the rational use of land. Therefore, the regional authorities strengthened the state land supervision, in order to identify idle agricultural land.

Postagrogenic lands are a rather heterogeneous group of lands that differ in floristic composition, species saturation, horizontal and vertical structures, dominant species, productivity, economic value and in soil characteristics (structure, water regime, soil fertility, etc.). Some of these lands with low-power horizon as a result of ill-conceived plowing were exposed to wind erosion. Now they belong to unproductive soils and therefore can't be used as arable land. All this leads to the need to make a decision on the further use of postagrogenic lands in agriculture (for example as pastures and hayfields) or to exclude their agricultural turnover for other purposes (creation of forest plantations, allocation for industrial and residential areas, etc.). Otherwise, there is a great risk of deterioration of their quality and the development of negative processes. Overgrowth of fallow lands with tree-shrub vegetation is noted as one of the negative factors. From the forestry point of view – this process is natural and corresponds to the succession development.

Quite a lot of works of Russian and foreign researchers are devoted to the study of tree invasion on old arable lands. The study of the space-time evolution of postagrogenic lands and assessment of their overgrowth, were carried out by many authors, but mainly in relation to the forest zone of the European part of Russia, where abandoned agricultural lands are overgrown with quite valuable small-leaved and coniferous tree-shrub species. Geobotanists and foresters described the stages of succession dynamics and the composition and productivity of the emerging forest plantations [1-4]. In the steppe and forest-steppe zones of Siberia, invasions of tree species (*Populus nigra L., P. laurifolia Ldb., Ulmus pumila L., Betula pendula Roth, Larix sibirica Ledeb*) are described on postagrogenic lands. Most stands formed naturally on former agricultural land are characterized by high productivity and practically do not differ from plantations created by artificial means. However, in some regions of Russia, including Khakassia, the fallows are overgrown with low-value tree Elm Siberian (*Ulmus pumila L.)* [5-6]. There are no natural elm community on the territory of Khakassia and South of Krasnoyarsk region. During the renewal of *U. pumila* on postagrogenic lands the source of seeds act as protective forest belts. Elm Siberian has always been considered a promising breed in the creation of protective forest strips. It is drought-resistant, exists on dry soils, withstands significant salinization of soils, well resumed self-seeding [7].

Cause of the best microclimate (shade for grazed animals) and more favorable conditions for the growth of various types of grasses and the emergence of ecological niches, such as habitats of steppe animals, birds and useful entomofauna are formed, a number of authors believe that the agroforestry ecosystems formed in such cases significantly increase the productivity of such lands as pastures. [8-9]. However, if it is an uncontrolled human process, such areas are sources of «biological pollution» and the object of increased fire hazard (lack of care contributes to the accumulation of dead wood). The resulting stand slows down and alters the process of demutation, and also makes it difficult to use the fallow for mowing and grazing cattle [5].

The purpose of our work is to assess the geoecological state of prostagrogenic lands, to identify the share of the total area of agricultural land of the Moscow village council of Ust-Abakan district of the Republic of Khakassia and to determine their stage of overgrowth. To do this, on the basis of ground and satellite data it's necessary to create a GIS project «Geoecological assessment of postagrogenic lands of the Moscow village council» containing a series of maps reflecting the current state of postagrogenic lands.

The characteristic of the processes of overgrowth of agricultural land with tree-shrub vegetation was carried out on the territory of the Moscow village council of Ust-Abakan district of the Republic of Khakassia. The study area with a total area of 54400 hectares is located in the North-Eastern part of Ust-Abakan district and occupies 7,2 % of its total area. From the Eastern part of Moskovsky village council borders with the Krasnoyarsk reservoir, from the North with Bogradsky district, from the South it borders with the urban district of Chernogorsk and Sunny selsoviet, from the Western part it borders with Vershina-Bijinsky selsoviet. On its territory there are 3 settlements with a total population of 1783 people with an administrative center in the village of Moskovskoye (Fig. 1) [10].

The relief of the studied territory is a weakly hilly surface with alternation of low alluvial plains with steep and ridge denudation increases, monoclinal Cuesta ridges with sharply asymmetrical slopes and wide flat hollows between ridges with an absolute height of 300 to 500 m above sea level. The Duration of the period with an average daily air temperature above 10° is 110-120 days. The sum of the average daily air temperatures above 10° - 1700-1900°. The duration of the frost-free period varies from 105 to 115 days. The amount of precipitation for the year is 310-340 mm. the amount of precipitation for may-July generally does not exceed 170 mm. The dryness Index is more than 1.9 (dry) [11].

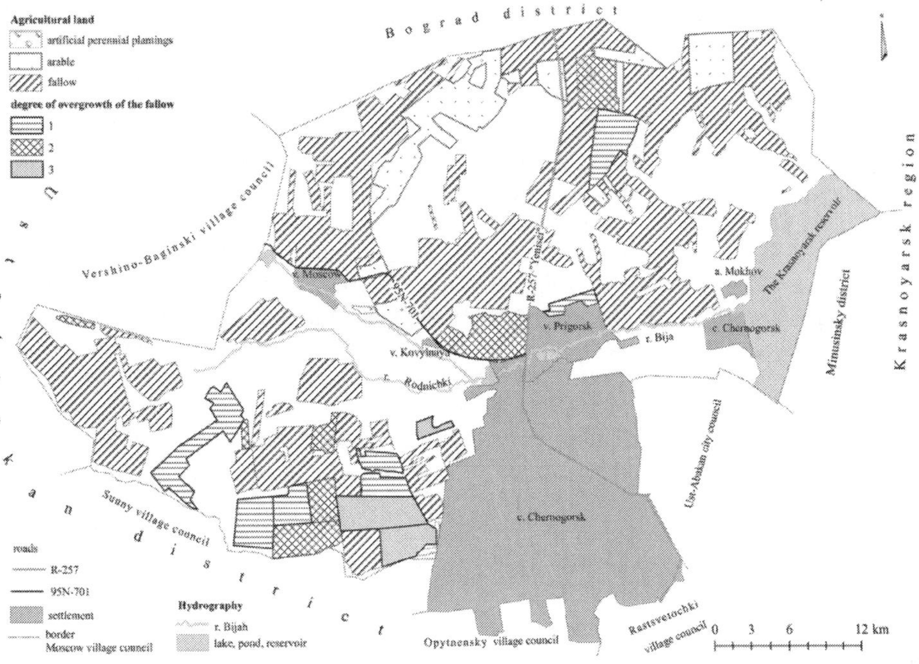

Fig. 1. The map of overgrowth of agricultural land with tree-shrub vegetation on the territory of the Moscow village council of Ust-Abakan district

From West to East, the village council is crossed by the river Bija, which flows into the Krasnoyarsk. It is the left tributary of the reservoir. The river Rodnichki collects water from the North-Western part of the village council and is a right tributary of the river Bija (flows into the river in the area between the settlements Kovylnaya and Prigorsk). A system of ponds for irrigation, watering of pastures, irrigation fields and artificial perennial plantations in the area of the urban-type settlement of Prigorsk was created on the Bijah river. There are also small temporary streams and streams on the logs and also lakes with an area of less than 1 km^2 are usually dry in summer. There are wetlands in the area of farms number 18 and number 11, as well as in the valley of the river Bijah.

Soils are mainly consist of ordinary chernozems and chestnut soils, often saline due to the close occurrence of mineralized groundwater and saline parent rocks. Saline soils in some areas make up even the main background. The main type of vegetation is low-production small-bunchgrass steppes with a high proportion of sagebrushes.

The basic material for the work were the analysis of cartographic materials (agricultural carat Ust-Abakan district, a series of thematic maps for planning of agriculture in the south of the Krasnoyarsk region and others), the results of their ground observations (composition and structure of vegetation postagrogenic lands (floristic composition, projective cover of grass, the ratio of botanical groups, height and age of tree-shrub bushes, etc.), data, public land maps and multi-temporal remote sensing data.

Esri's ArcMap program was used to create a GIS project for the overgrowth of postagrogenic areas. In the GIS project, the contours of agricultural territories were digitized on the basis of the agricultural map of Ust-Abakan district. Esri's World Imagery base layer was used to highlight different types of areas (https://services.arcgisonline.com/ArcGIS/rest/services/World_Imagery/MapServer). Also, to clarify the areas and the degree of overgrowth were deciphered at different times (since 2010 to 2019) satellite images of Landsat 4-5, 7, 8 and Sentinel 1 and 2 taken from the archives of the United States Geological survey (USGS, https://earthexplorer.usgs.gov). The ArcMap work project settings are as follows: WGS 1984 UTM zone 46N coordinate system, Mercatora / WGS84 projection. According to the results of the analysis of attribute information, calculations of arable land areas and postagrogenic areas with different degrees of overgrowth were made.

The Moscow village council in its structure belongs to the agro-industrial complex. On the Moscow village council's territory the share of agricultural land accounts for 90.7 % of its total area. The economy of the Moscow settlement is mainly based on the conduct of agricultural production of peasant farms and personal subsidiary farms. The most of the land is represented by fallows since the area is in the zone of risky agriculture and the soil with poor low-spent fertility. The main activity of the peasant farm is the breeding of sheep and cattle. Only 2 entrepreneurs are engaged in growing grain and leguminous crops. The total area occupied by arable land is only 7761,6 hectares. These lands are located on the border with the Bograd district where the greatest amount of precipitation falls and soils are more fertile. Most of the agricultural land set aside for arable land and previously used for growing crops is now in various stages of recovery. As a rule, cause of the lack of productive hayfields and pastures (mainly small-bunchgrass steppes with a high proportion of sagebrush), fallows are used for hay harvesting and cattle grazing.

On the lands of the Moscow village council a forest reclamation is partially implemented. Because of the lack of moisture in the territory, *Ulmus pumila L.* was used in conjunction with *Caragana arborescens Lam* to create protective and pasture-protecting forest strips. Forest reclamation was carried out on three sections of the village council: in the southern part of the village council №. 23, in the area of farm; between Kovylnaya and Prigorsk in the central and the Northern parts of the village council on the border with Bogradsky district. Forest belts are mainly 3 rowed of the blown type. The direction of the forest belts from North to South, across the prevailing winds. In total, there are 187 bands with varying degrees of damage (completely lost bands about 35 %).

Existing methods of score assessment of the current state of agricultural land exposed to overgrowth processes can't be used for the territory of Khakassia, as the forming stand of elm Siberian is strikingly different from the forests (coniferous, small-leaved), formed in other regions of Russia. Therefore, in relation to the territory of Khakassia, based on the analysis of long-term field data, a score assessment of the current state of deposits subject to overgrowth processes was developed. The following stages of the processes of overgrowth of agricultural land with tree and shrub vegetation were identified:

1. Zero stage of overgrowth – there are no overgrowth processes.

2. The first stage of overgrowth – there is an undergrowth of tree species commensurate with the grassy tier with a projective cover up to 50-100% especially near the mother plantings (forest belts).

3. The second stage of overgrowth - begins thinning of the undergrowth due to the biological characteristics of elm (a large percentage of freezing plants), the completeness of the stand is reduced to 0,3-0,5.

4. The third stage of overgrowth is further thinning of the stand. The completeness is very low $\leq$ 0.1. The average height of the elm is at least 2 m (the age of the plantings is usually more than 8 years). Again there is a young sapling.

According to the data obtained, 29,5% of the land from the total area of the Moscow village council is accounted for by a fallow. 77 % of the land (15432 ha) are subject to overgrowth processes, 43 % are in the first stage of overgrowth, 38,3% in the second and 18,7 % in the third. Haymaking and grazing on fallow lands constrain the appearance of woody vegetation, but if they stop *Ulmus pumila L.* for several years actively spreads over abandoned fields.

Protective forest strips play a multi-functional positive role for agricultural fields. However, in the absence of proper care for forest belts and a spontaneous conservation of land between the forest belts, overgrowing of fallow fields with elm tree and shrub vegetation occurs. Geoecological assessment of the features of overgrowth and growth of tree-shrub plantations on postagrogenic lands, remote sensing data and geoinformation technologies will provide not only new knowledge about their space-time evolution organization but will also provide a basis for the formation of a science-based system of measures for rational agriculture.

References

1. S. V. Zalesov, N. N. Novoselova, L. P. Abramova, Forests of the Urals and the economy, vol. **25**, pp. 30 – 41 (2004)
2. E. V. Belorussia, Modern problems of remote sensing of the Earth from space. vol. **9**, № 1. pp. 57-64 (2012)
3. V. V. Belyaev, O. D. Kononov, A. A. Karaban, V. V. Staritsyn, Bulletin of the Northern (Arctic) Federal University. Series: Natural Sciences. №. **2**., pp. 5-11 (2013)
4. L. V. Golubeva Forest-ecological transformation of postagrogenic lands on carbonate deposits in the subzone of the middle taiga of the Arkhangelsk region: dis.... Cand. of agricultural Sciences. (Arkhangelsk, 2015)
5. S. A. Holboeva, A. B. Sakhyaeva, R. Yu. Abasheev, Bulletin of NEFU, № **6** (**68**), pp. 43-54 (2018)
6. M. A. Martynova, A. I. Lobanov, Problems of botany in southern Siberia and Mongolia: collection of scientific articles based on the materials of the XIV international scientific and practical conference (may 25-29, 2015, Barnaul), pp. 205-208 (2015)
7. N. Likhovid, Forest belts in Khakassia, p. 47 (1969)
8. V. K. Savostyanov, Achievements of science and technology of agriculture, №. **4**, pp. 4-9 (2011)
9. E. N. Savin, N. I. Likhovid, S. M. Charkov, I. I. Glukhov, Technology of creation of protective forest stands for the purposes of animal husbandry in the steppe regions of Khakassia, p.13 (1995)
10. D. S. Savin, Scheme of territorial planning Of the municipality of Ust-Abakan district of the Republic of Khakassia. Regulations on territorial planning of the municipality of Ust-Abakan district of the Republic of Khakassia, vol. **1**, p. 29 (2012)
11. South of the Krasnoyarsk Territory. Series of maps for agricultural planning, 39 p. (1977)
12. E. A. Demina, Ecology of Southern Siberia and adjacent territories, vol. 2. Issue **23** pp. 15-16 (2019)

https://doi.org/10.1051/e3sconf/202014903007

Variations of atmospheric aerosol parameters in periods of seismic activity in Tien-Shan

Leonid Sverdlik[1,*]

[1]Research Station of the Russian Academy of Sciences in Bishkek, Kyrgyzstan

Abstract. Combination of numerical modeling (MERRA) and ground-based (AERONET) monitoring of atmospheric aerosol parameters were used to discover a possible connection to seismic activities within the territory of Tien-Shan. The obtained results demonstrated consistent behavior between increase of aerosol optical thickness (AOT) and powerful earthquakes, epicenters of which were located within 200 km from the "Issyk-Kul" station (AERONET). Thereat, the dominant role was played by fine-dispersed particles, which virtually determined the nature of AOT evolution. We also observed well-defined changes of the ratio between mass fraction $PM_{1.0}$ (particles less than 1.0 μm aerodynamic diameter) and optical thickness of aerosol scattering several days before an earthquake of M > 5.0 magnitude.

1 Introduction

By now numerous anomalies, observed before seismic events both on the Earth's surface and in the atmosphere and ionosphere, have been detected while a reliable earthquake forecast is still one of the most important unsolved geophysics problems, which assumes a necessity to investigate physical characteristics of all the complex of phenomena in geospheres. An important aspect from the point of view of understanding of lithospheric-atmospheric interrelation during seismic activities is investigation of spatial and temporal variations of atmospheric parameters and atmosphere aerosol characteristics in particular [1]. Aerosols play a significant part in the Earth's radiation balance [2], regional and global climate changes and the also influence on air quality and public health. In a series of researches particulate matter optical properties were used as indicators of pre-seismic anomalies [3]. Thereat, changes of atmospheric aerosol optical thickness (AOT) are analyzed more often. Thus, assessments of AOT changes, occurred before strong earthquakes in Chile (2010) and in Nepal (2015), showed an express increase (approximately by ~40%) in vicinity of the epicenter several days before the main event [4]. Anomalies in AOT temporal and spatial variations were observed before, during and after each of eight strong earthquakes of M > 8.0 [5]. Investigations of relationship between large seismic events and anomalous variations of aerosol optical parameters are based on data of both land and spaceborne remote sensing

* Corresponding author: l.sverdlik@mail.ru

[1]. The most accurate information on atmospheric aerosol parameters are obtained due to measurements with CIMEL sun photometers, which are used in the global ground-based aerosol observation network, AERONET [6]. The data may be effectually used in seismically active regions in vicinity of AERONET stations. In recent years, satellite measurements, allowing analysis of atmospheric properties changes on a global basis, have become widely used to monitor all types of hazardous natural phenomena. Combination of the two information sources is most effective and, accordingly, it was used in this study. For the quantitative assessment of the aerosol content in the atmosphere, results of ground-based measurements of spectral AOT before and after earthquakes were used. Satellite time series of various combinations of parameters and particularly those providing important information in relation to the ratio between aerosol scattering properties and the mass fraction of particulate matter with aerodynamic diameter of less than 1.0 μm ($PM_{1.0}$).

2 Initial data

Ground-based and satellite data for periods of preparation of 12 strong earthquakes (EQ) characterized by magnitudes of M > 5.0 occurred within the territory limited by the following coordinates: 39–44° N, 70–80° E in 2003–2018 have been analyzed (Table 1). Information on seismic events were obtained from the earthquake catalog composed on the basis of data of the KNET regional seismological network [7] and the online version of the U.S. Geological Survey (USGS) global catalog [8].

Table 1. Characteristics of the study sample of seismic events (M > 5.0).

EQ	Date	Time	Latitude	Longitude	Depth, km	M
01	22.05.2003	18:11:56	42.95° N	72.76° E	33.0	5.5
02	16.01.2004	09:06:21	42.53° N	75.35° E	27.5	5.0
03	08.01.2007	17:21:50	39.80° N	70.31° E	16.0	6.0
04	17.03.2011	15:34:35	40.48° N	79.10° E	32.8	5.0
05	09.04.2011	12:30:26	42.07° N	74.82° E	14.4	5.0
06	28.01.2013	16:38:54	42.61° N	79.71° E	15.0	6.1
07	23.11.2013	09:42:08	42.53° N	75.71° E	9.8	5.0
08	14.11.2014	01:24:17	42.09° N	77.26° E	10.1	5.2
09	17.11.2015	17:29:36	40.38° N	73.20° E	18.0	5.6
10	19.07.2016	12:32:56	41.55° N	79.01° E	10.0	5.1
11	26.03.2018	12:53:50	43.72° N	77.33° E	29.0	5.0
12	14.09.2018	22:15:03	41.85° N	77.13° E	18.0	5.0

Initial data on atmospheric aerosol parameters were obtained at the "Issyk-Kul" station, which is a part of the AERONET global network [9], the data are represented by AOT daily means of reliable quality (Level 2.0), as well as data of Level 1.5 (cloud screening), that have a higher error of atmosphere optical parameters reconstruction, but provide a better measurement statistics [10]. Satellite remote sensing data, processed in the MERRA model (Time Series; Area-Averaged of Dust Scattering AOT 550 nm – PM 1.0 μm; M2T1NXADG v5.12.4) were used as a means of global monitoring of variability of aerosol characteristics [11], in this particular case, optical thickness of aerosol scattering at the wavelength 550 nm for mass concentration of fine particles $PM_{1.0}$, were used. Both time series of the parameter with discretization of $\Delta t = 1$ hour above areas of ±1° in size from earthquake epicenters, and its spatial distribution in the area of the seismic focus of each event [12] were recovered.

3 Consideration of the results

3.1 Aerosol parameters variations according to land measurement data

Investigation of possible relationship between the atmosphere aerosol load and seismic activities in Tien-Shan was performed using land observation data. For this purpose the strongest seismic events, epicenters of which were located within ~200 km from the AERONET station were selected. One of that kind of earthquakes of magnitude 6.1 was registered on January 28, 2013 on the border between Kyrgyzstan and Kazakhstan (Fig. 1a). Fig. 1b illustrates AOT (Level 1.5) variations in the spectral range from 340 to 1020 nm in periods from January 4 to February 4, 2013. As it appears from the presented diagrams, the aerosol optical thickness in the short-wave region of the spectrum (340–500 nm) achieved maximum values (0.400–0.550) 1–6 days before the seismic event.

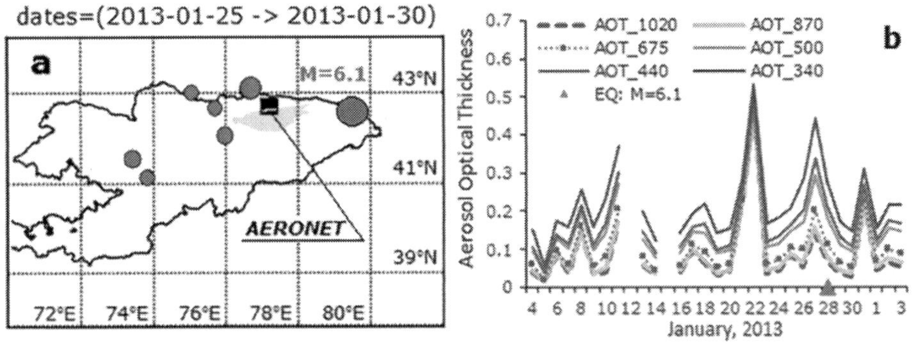

Fig. 1. Locations of the AERONET station (42.623 N, 76.983 E) and earthquake epicenters (a) and change of daily means of spectral AOT (Level 1.5) in January 2013 (b). The marker shows the earthquake of magnitude M = 6.1 (January 28, 2013).

AOT observation data, adjusted according to results of recalibration (Level 2.0), were detailed using fine (AOT_F) and coarse (AOT_C) modal constituents. The sum of the fine- and coarse-mode AOTs constitutes the total AOT. As is shown in Fig. 2, the highest value of the optical thickness (0.325) at the wavelength of 440 nm was observed a day before an earthquake of magnitude 6.1.

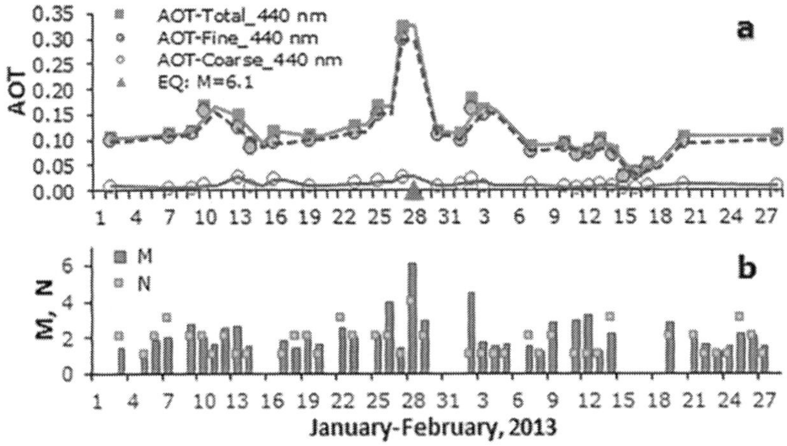

Fig. 2. Variations of AOT, AOT_F and AOT_C (Level 2.0) daily means with time at the wavelength of 440 nm (a) and seismic activities (b) in January–February 2013.

Assessments of the contribution of two aerosol fractions demonstrate that the dominant role was played by fine-dispersed particles, which virtually determined the nature of the AOT evolution. Apparent consistency of behavior between AOT_F variations (Fig. 2a) and the seismicity, which is represented by a sequence of magnitudes (M) and the number of events (N) (Fig. 2b), allows to make an assumption as to possible sources of aerosol particles. One a the reasons of finely dispersed aerosol domination in the atmosphere may be nucleation processes emerging in the course of ionization of near-ground atmospheric layer and planetary boundary layer [7].

Another example of photometric measurements in periods of preparation and occurrence of earthquakes of magnitude $M > 5.0$ (September 3 and 14, 2018) may be results of calculation of AOT (Level 1.5) daily means at the wavelength of 340 nm and corresponding variations of fine mode fraction (FMF) contribution into the aerosol optical thickness in August-September 2018 (Fig. 3a). As we can see, in periods preceding strong earthquakes changes in particle size distribution of atmospheric aerosol were observed. Increased AOT values were generally determined by fine-dispersed particles (Fig. 3b).

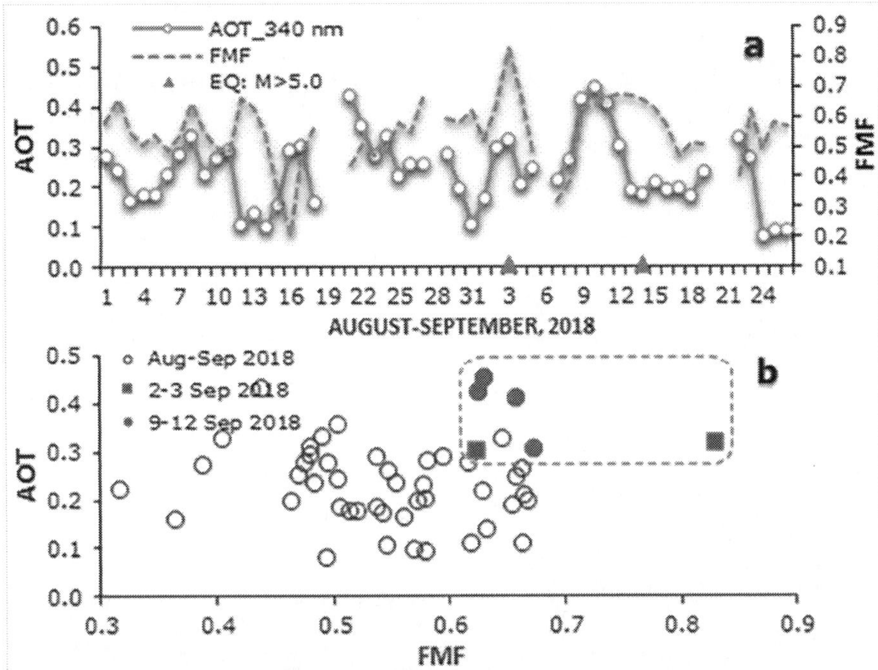

Fig. 3. Time series of AOT (Level 1.5) daily means at the wavelength of 340 nm and fine mode fraction (FMF) relative contribution into the aerosol optical thickness in August-September 2018 (a). The markers indicate earthquakes of M > 5.0 (September 3 and 14, 2018). Interrelation between AOT and FMF in periods of seismic activities (b).

3.2 Results of satellite measurements of aerosol parameters

Processes, taking place in geospheres in the course of preparation of strong earthquakes, may result in formation of aerosols, size distribution of which differs from that observed in periods without seismic events. Such peculiarities of particle size spectra pre-seismic variations were detected using the $\mu = AOT / PM_{1.0}$ parameter, characterizing relation between mass concentration and optical thickness of aerosol scattering [8]. Preliminary results of satellite data processing in the form of normalized ratios $1/\mu$, normalized to one earthquake moment, are presented in Fig. 4. Apparent well-defined changes $1/\mu$, preceding strong earthquakes of

M ≥ 5.0 by 1–5 days may be probably attributed to presence of predominantly fine-dispersed particles, which have a great impact on AOT, but insignificantly change $PM_{1.0}$.

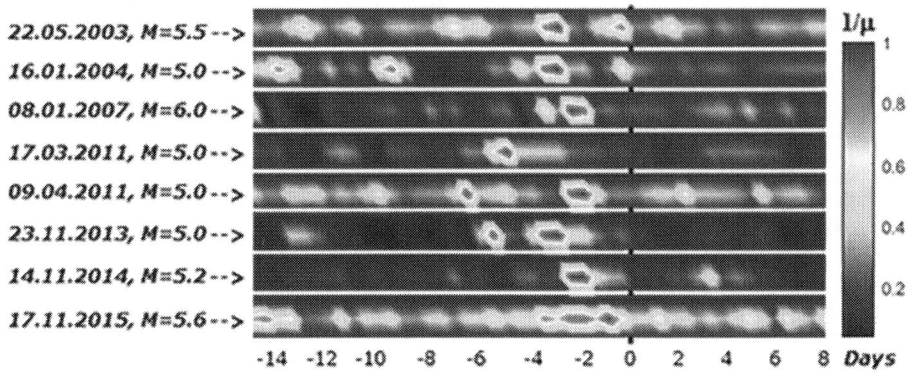

Fig. 4. Distribution of normalized ratios 1/μ in periods starting on 15 days before and ending 7 days after each event M > 5.0 in Tien-Shan.

Spatial distribution of the Dust Scattering $AOT_{550\ nm} - PM_{1.0\ \mu m}$ parameter a day before strong earthquakes, occurred on November 17, 2015 (M = 5.6), July 19, 2016 (M = 5.1), March 26, 2018 (M = 5.0) and September 14, 2018 (M = 5.0), is presented in Fig. 5. Local maximums of 1/μ were observed near epicenters one day before earthquakes.

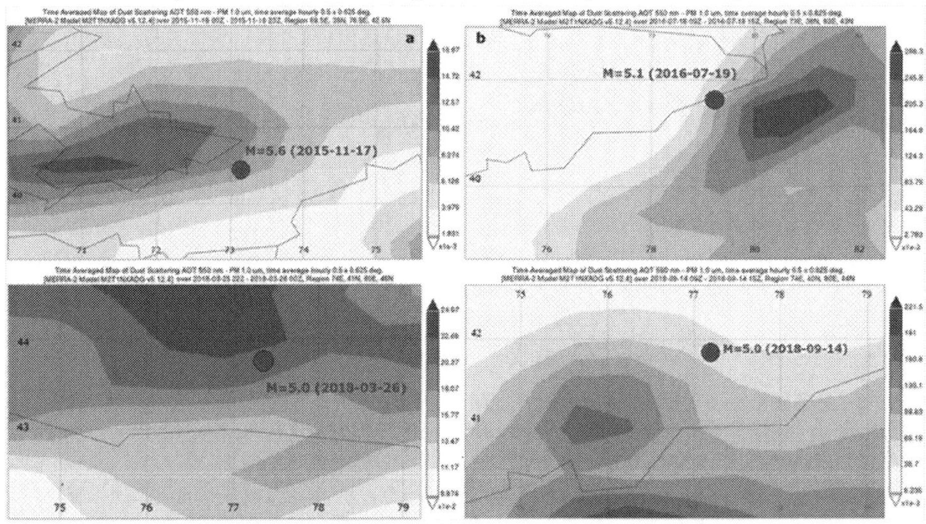

Fig. 5. Spatial distribution of Scattering $AOT_{550\ nm} - PM_{1.0\ \mu m}$ one day before strong earthquakes of M > 5.0: November 17, 2015 (a); July 19, 2016 (b); March 26, 2018 (c) and September 14, 2018 (d).

5 Conclusion

Results of land (AERONET) and satellite (MERRA) measurements of atmospheric aerosol parameters against several events of seismic activity in the territory of Tien-Shan have been analyzed. Periods preceding strong earthquakes, epicenters of which were located within ~200 km from the AERONET station have been examined. Analysis of ground-based remote sensing demonstrated apparent anomalous changes of the aerosol optical thickness (AOT > 0.3) several days before the earthquake. Thereat, dominating contribution to increased values of the aerosol optical thickness was made by fine-dispersed particles (FMF > 0.6), which virtually determined the nature of AOT evolution. The changes of the aerosol characteristics may be associated with seismic events.

An important aspect from the point of view of understanding of lithospheric-atmospheric interrelation during seismic activities is investigation of not only temporal variations, but spatial distribution of aerosol characteristics as well. Atmospheric effects accompanying processes of preparation of strong earthquakes in the form of change of the ratio of aerosol scattering optical thickness at the wavelength of 550 nm to mass fraction of particulate matter with aerodynamic diameter of less than 1.0 μm ($PM_{1.0}$). Spatial distribution of the parameter was characterized by presence of local maximums observed one day before strong earthquakes of $M > 5.0$ near epicenters of the events under examination.

The above examples confirm presence of characteristic features in the temporal dynamics of atmospheric aerosol parameters in pre-seismic period. However, the results obtained are only preliminary and require a more detailed investigation.

Acknowledgements

The author is grateful to Giovanni team of NASA GES DISC for free access to satellite measurement data and modeling results. I also thank the NASA Goddard AERONET team led by Dr. B. Hoblen and station manager M. Orozaliev for producing surface observations at Issyk-Kul station.

The study is partially performed within the framework of the state task of the Federal State Budgetary Institution of Science, Research Station of Russian Academy of Sciences in Bishkek (theme No. AAAA-A19-119020190064-9).

References

1. Q. Liu, X. Shen, J. Zhang, J. Cui, S. Zhao, Geophysical Research Abstracts. **21**, 3938 (2019).
2. L.G. Sverdlik, Bulletin of the Kyrgyz-Russian Slavic University. **12**, 156 (2018).
3. Z-H. Jiao, J. Zhao, X. Shan, Nat. Hazards Earth Syst. Sci. **4**, 1013 (2018).
4. S.A. Pulinets, D.P. Ouzounov, D.V. Davidenko, Moscow. Trovant Publ. 144 (2014).
5. M. Akhoondzadeh, F.J. Chehrebargh, Advances in Space Research. **58**, 890 (2016).
6. D.M. Giles, A. Sinyuk, M.G. Sorokin, J.S. Schafer, A Smirnov., I. Slutsker, T.F. Eck, B.N. Holben, J.R. Lewis, J.R. Campbell, E.J. Welton, S.V. Korkin, A.I. Lyapustin, Atmos. Meas. Tech.. **12**, 169 (2019).
7. Earthquakes catalog of Institute of Seismology of NAS KR and Research Station of RAS in Bishkek.
8. URL: https://earthquake.usgs.gov/earthquakes/.
9. URL: https://aeronet.gsfc.nasa.gov/.
10. L.G. Sverdlik, Science, new technologies and innovations. **3**, 3 (2013).
11. C.A. Randles, A.M. Da Silva, V. Buchard, P.R. Colarco, A. Darmenov, R. Govindaraju, A. Smirnov, B. Holben, R. Ferrare, J. Hair, Y. Shinozuka, C.J. Flynn, J. Clim. 30. **17**, 6823 (2017).
12. URL to Reproduce Results: http://giovanni.gsfc.nasa.gov/.

DYNAMICS OF POST-FIRE EFFECTS IN LARCH FORESTS OF CENTRAL SIBERIA BASED ON SATELLITE DATA

Nikita Yakimov[1,3], and *Evgenii* Ponomarev[1,2]

[1]Federal Research Center "Krasnoyarsk Science Center of the Siberian Branch of the Russian Academy of Sciences", Krasnoyarsk, 660036, Russian Federation
[2]Sukachev Institute of Forest, Siberian Branch, Russian Academy of Sciences, Krasnoyarsk, 660036, Russian Federation
[3]Siberian Federal University, Krasnoyarsk, 660041, Russian Federation

Abstract. The article represents the results of Terra, Aqua / MODIS, Landsat-8/OLI satellite data analysis for fire damaged plots in larch forests of Central Siberia. The analysis of averaged surface temperature (brightness temperature) and vegetation index (NDVI) was performed for post-fire circumstances. Estimates of the state and dynamics of fire-damaged vegetation cover were obtained on the basis of inter-seasonal variation of the NDVI index. It was found that post-fire dynamics of vegetation cover determines the surface temperature anomalies within the fire scar plots during at least five years after wildfire impact. It was instrumentally registered that the maximum excess of brightness temperature on post-fire areas can reach up to 11°C comparing to that of background areas under the same conditions. Such anomalies are determined by higher level of insolation due to partial or total tree mortality, as well as by decreasing of on-ground cover thickness after fire impact on grass and moss-lichen covers. During the first year after a fire in larch forests of Siberia, the maximum temperature anomalies of the underlying surface was recorded in the third decade of June. In the course of 2–5 years after burning, the maximum temperature anomalies shift to the second or even third decade of July within the phenological season. The suggested approach allows to assess the degree of fire impact on vegetation, as well as to predict changes in the active layer of permafrost soils, which may be a consequence of extra thermal flow at the surface in the circumstances of disturbed larch forests of Siberia.

1 Introduction

Since the end of XX–begin of XXI в. there is a trend of growth in the number and area of wildfires, which determines the increase in the percentage of disturbed areas in larch forests of Central Siberia [1–5]. Boreal forests around the world are characterized by high levels of fire activity [6–7]. In Siberia, among others, the long-term cumulative effect of fire impact is expressed in the percentage of forest areas covered by fire, which reaches 20–25% over the past 20 years [5; 8] of the total area of larch forests. Due to the impact of wildfires there are anomalies of the thermal regime of the underlying surface, which observed for a long time within disturbed forest areas [8]. Such changes are caused by a decreasing of albedo and heat-insulating properties of the litter and ground cover at the post-fire forest plots [9; 10; 11]. Anomalies of the thermal balance of the underlying surface can be critical for ecosystems under the conditions of the permafrost of Central Siberia [12], as an ecosystem under long-term disorders are characterized by low stability. The ecological importance of Northern forests prevails over their raw material potential [9], and the impact of wildfires can lead to significant changes in environmental conditions.

Nowadays long-term monitoring of the fire impact on the forests of Siberia cannot be carried out without the use of satellite monitoring tools [13; 14]. Thematic processing of multispectral satellite data makes it possible to identify post-fire changes, as well as to evaluate their intraseasonal and interseasonal dynamics of burned areas Thus, modern approaches based on remote sensing data make it possible to control and predict the state and succession processes in forest ecosystems.

The purpose of this work – quantitative assessment of post-fire changes in larch forests, which describing the state of vegetation and ground cover during 5 years of reforestation. These data are necessary for modeling of thermal processes dynamics in the upper soil horizons and forecasting the state of seasonally thawing layer of soil under permafrost condition. Increasing areas of wildfire damage actualizes research in this direction around the world [15; 16]. For the territory of Siberia, data on temperature anomalies of the underlying surface after logging are presented [17]. The effects of long-term temperature anomalies on soils and dynamics of intra-seasonal groundwater flow of rivers were recorded [18–20]. In this paper, the following aspects are considered: absolute and relative (comparing to background plots values) anomalies of temperature and NDVI; their intra-seasonal dynamics; as well as trends of recovery of NDVI values and averaged brightness temperature of underlying surface under the conditions of 5 years post-fire restoration. The analysis summarizes the data obtained from satellite surveys, mainly for post-fire areas in the larch forests of the permafrost zone of Central Siberia (Evenkia).

2 Initial data

The materials of the satellite survey Terra/MODIS (Moderate Resolution Imaging Spectroradiometer) with a spatial resolution of 250-1000 m were used. Standard MODIS products of the L2 and L3 processing level were used (available on the USGS website, URL: https://lpdaac.usgs.gov/dataset_discovery/modis). The total surface albedo information was calibrated in the ranges $\lambda 1$ = 0.620-0.670 μm (channel 1, MODIS) and $\lambda 2$ = 0.841-0.876 μm (channel 2, MODIS) obtained from the standard product MOD09GQ. Daily surface temperature data in the range $\lambda 3$ = 10,780-11,280 μm (31 channels, MODIS) were calculated from the product MOD11A1. Landsat-8/OLI satellite materials with a spatial resolution of 15-30 m (free available on the USGS website, URL: https://earthexplorer.usgs.gov) were used to compare the boundaries of wildfires.

The analysis of values averaged over 7-12 post-fire sites over a five-year period was performed. The number of plots were depended on the year and availability of informative images for phenology decades of the season. The intra-seasonal variation of the studied parameters was analyzed for the growing season taking into account the phenological cycle of vegetation, the characteristic duration of the individual phases of which is about a decade.

The geographical coordinates of the site on the territory of Central Siberia: 59.8-62.5°N and 96.5-104°E. The length of the research area from North to South is 330 km, and from West to East is 400 km. The district is located on the Central Siberian highland, the terrain is mainly mountainous, disturbed areas of the forest are represented by post-fire larch with dwarf shrub, lichen and mosses.

Anomalies of NDVI and temperature field were defined both in absolute values in comparison with background values, and in relative indicators (percent in relation to average long-term norm).

3 Methods and data processing algorithms

Within the boundaries of the area of interest, a selection of retrospective survey materials Terra, Aqua/MODIS, Landsat-8/OLI for the period 2013-2018 was carried out. On the selected images wildfires of different years were detected, Dating was carried out using the database of fires Sukachev Institute of Forest Siberian Branch, Russian Academy of Sciences, containing materials during the period of observation 1996-2019 years. Local areas were considered, in which an average of 10 or more large fires were detected. Analysis of local wildfires made it possible to increase the information content of the image during the sub-stage examination. We selected images of this area for each of the decades of the season with maximum information. For each decade of the season, averaged values of brightness temperature of disturbed areas (31 channels) and surface albedo (1, 2 channels) were obtained. Background values were recorded in the immediate vicinity of post-fire areas at a distance of 1-5 km in similar forest conditions. An example of the distribution of parameter measurement points is shown in figure 1.

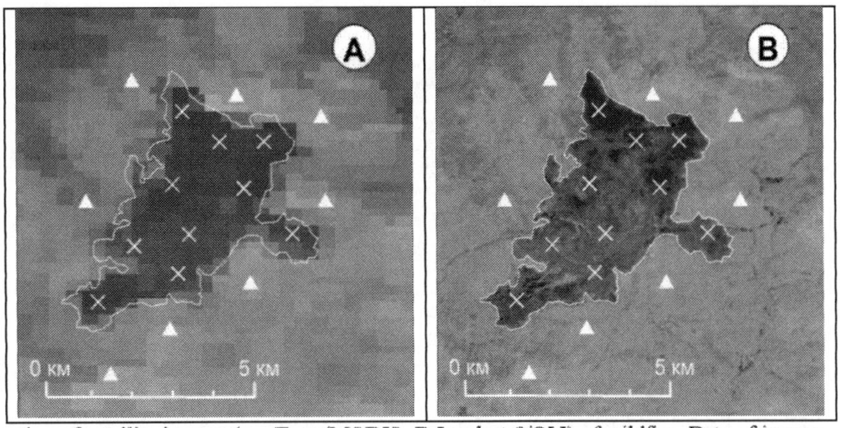

Fig. 1. Illustration of satellite images (a – Terra/MODIS, B-Landsat-8/OLI) of wildfire. Date of images – 23.06.2018. Conventional marks indicate the points of selection of spectral data on the brightness temperature and albedo of the underlying surface inside the fire range (X) and background values in the immediate vicinity (▲). Central Siberia, Evenkia, 2018

The obtained temperature values of the disturbed and background areas were applied to identify sub-cascade temperature anomalies. Albedo values were applied to calculate NDVI values averaged for each decade. On the basis of the obtained data, the dynamics of changes in the temperature anomaly and NDVI index in 1, 2, 4, 5 years after fires were constructed. There are no 2015 data defining the situation in the 3rd year after the fire, due to the low number of fires in the research area (2 fires) and in the territory of Evenkia as a whole.

4 Results and Discussion

In the first year after the wildfires, the temperature anomaly averaged 5°C for the summer and 4.6°C for the entire period of research (taking into account the third decade of May and the first decade of September). The maximum temperature recorded in the 3rd decade of June was 11°C (Fig.1/A). On average, the temperature in the 3rd decade of June was kept at 7.5°C. Such data are equivalent to the data obtained in other works [18, 19]. It has been suggested that the high level of insolation is caused by the oppression of the grass and moss-lichen covers and the low amount of tree mortality. In subsequent post-fire years, the maximum temperature values are shifted to mid-summer, 2-3 decade of July (Fig.1/B–D). This is confirmed by the data of field research [10]. In the second year after wildfires there is a decrease in the average temperature anomaly up to 40–47% depending on the decades. The average decline over the entire period is 27%. The maximum temperature anomaly was recorded in the 2nd decade of July and amounted to 7.64°C, which is 31% lower than the maximum temperature anomaly of the previous season. However, in the case of comparing the values of the second decade of July, this value decreased by only 15%. The values of the temperature anomaly with this decrease are 3.6°C for the summer period and 3.3°C for the whole period. In the fourth year after the fires, the temperature anomaly averaged 2.9°C for the summer and 2.7°C for the entire period. Thus, there was a decrease in the average temperature anomaly by 41% relative to the first year and by 19% relative to the second year. The maximum anomaly recorded in the 2nd decade of July – 5.34°C. In the fifth year of the recovery succession, the temperature anomaly decreased to 2.4°C over the summer and 2.3°C over the entire research period. The value of the temperature anomaly decreased by 50% relative to the first year, by 32% relative to the second year and by 15% relative to the fourth year. We observe a decrease in the temperature anomaly for five years, but during this time the difference in the temperatures of the disturbed and background values remains on average above 2±0.51°C. The Dynamics of the absolute temperature values is demonstrated in Fig. 2.

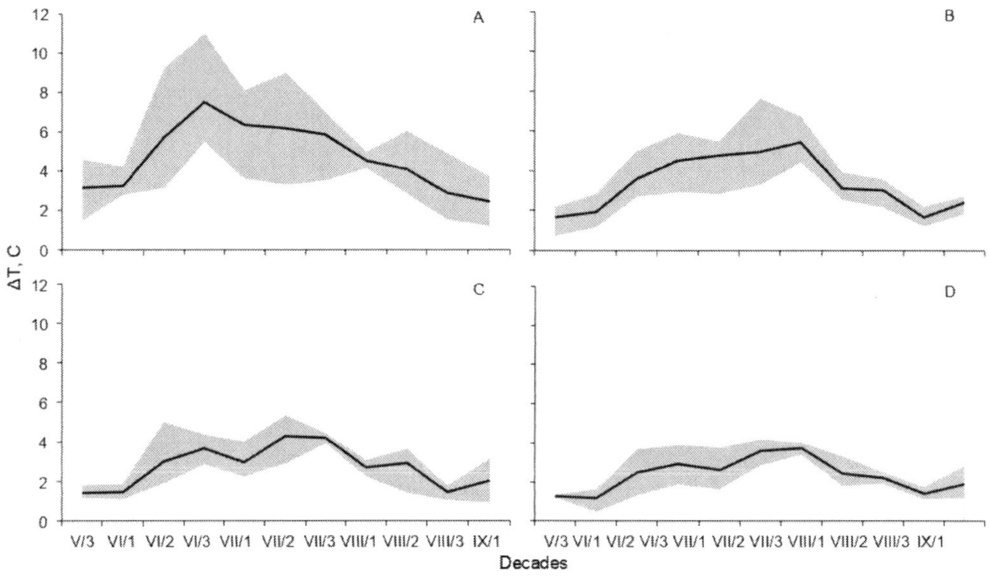

Fig. 2. Seasonal dynamics of absolute values of radio brightness temperature in the territory of Central Siberia (Evenkia) in relation to the background in 1(a), 2(B), 4(C), 5 (D) year after fires. On the x axis, the months/decades of the season are allocated: V-may, VI-June..., IX-September; 1, 2, 3-the name of the decades of months.

Based on the obtained values for the five years of recovery, it can be seen that the difference between the values of the temperature anomaly for the summer season and for the full period, including the values of the 3rd decade of May and the 1st decade of September, decreases and practically levels. This is due to the restoration of vegetation, which in turn smoothes the amplitude of the spread of minimum and maximum values. It is assumed that changes in the temperature anomaly by the sixth year of the recovery succession, differences in the temperatures of the disturbed areas, at the boundaries of the spring-summer and summer-autumn periods, will be at the level of error.

The whole season was divided into 3 time periods – early season (may–June), mid-season (July), late season (August–September). Thus, the obtained values of the relative brightness temperature anomaly in all post-fire successions showed a normal distribution of values with a maximum of temperature values in the middle of the season and low temperatures at the beginning and end of the season. The brightness temperature anomaly relative to the background values is shown in Table 1.

Tab 1. Relative anomaly of brightness temperature and NDVI

	Average anomaly of brightness temperature and NDVI, %							
	1 year		2 year		4 year		5 year	
	T_{rel}	NDVI	T_{rel}	NDVI	T_{rel}	NDVI	T_{rel}	NDVI
May-June	24,3±8,9	62,4±1,8	13,5±4,5	43,8±4,0	11,2±3,3	35,3±3,3	9,8±3,0	32,7±5,4
Jule	29,1±2,5	56,7±10,0	24,5±5,7	42,9±3,2	18,0±7,1	29,6±3,9	15,0±7,0	28,1±0,8
August-September	22,5±5,4	51,0±4,8	13,8±4,5	39,8±1,3	12,6±3,5	34,4±5,1	11,0±3,5	34,5±3,4

It is expected that the rate of decline of the temperature anomaly will continue, but less intensively, as the relative decline was 27% in the second year and fell to 15% by the fifth year, relative to previous years. Insolation of the underlying surface leads to a significant rise in surface temperature in the first year after the fire, but in subsequent years this value decreases. However, according to preliminary calculations of the temperature anomaly, at several fire stations in 2012, it reached 5°C, which is similar to the average for the first year.

NDVI values have also been declining annually for five years. The difference between the values for the entire research period and the values only for the summer period does not exceed 1%. Therefore, it is not necessary to divide the obtained values by the system (by the average for the summer period and the average for the whole period), as in the case of a temperature anomaly. The values of the relative anomaly of the vegetation index are presented in Table 1. The dynamics of the vegetation index NDVI is shown in Fig. 3.

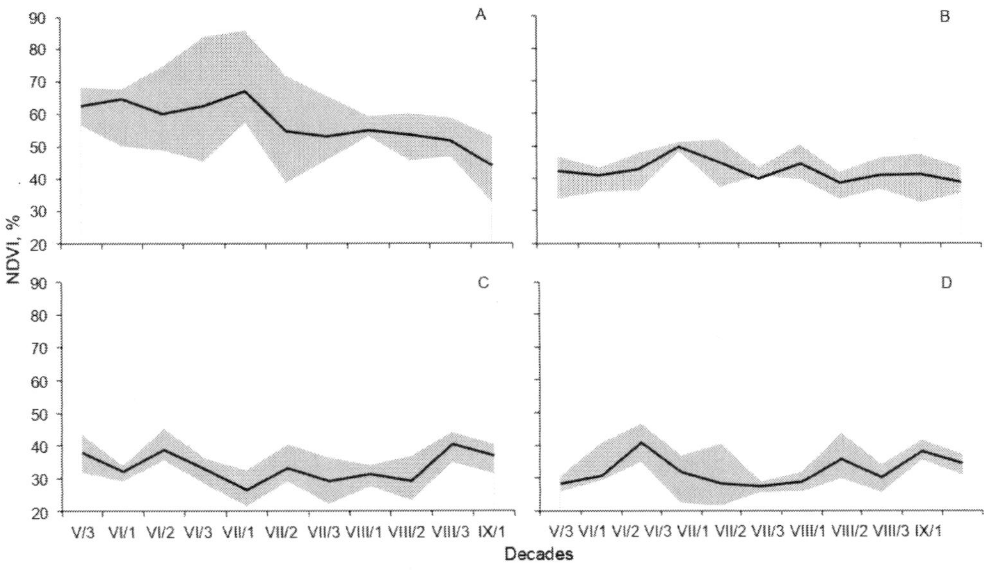

Fig. 3. Dynamics of vegetation index (NDVI) changes in 1(a), 2(B), 4(C), 5 (D) year after wildfires in Central Siberia (Evenkia). The dynamics of averaged values with limits in minimum and maximum values for each decade is illustrated. On the x axis, the months/decades of the season are allocated: V-may, VI-June..., IX-September; 1, 2, 3-the name of the decades of months.

In the first year after the fire, NDVI background values are higher than disturbed areas by an average of 56%, in some cases this value reaches 86% (in July), due to the high degree of oppression of vegetation cover in disturbed areas. In the second year, this anomaly decreased to 42%, and the maximum values reached 53%. Differences in indicators are smoothed out, consequently, there is a process of restoration of photosynthetic activity of vegetation. In the fourth year, this value of the vegetation anomaly decreased to 33%, and the maximum values were at the level of 44%. By the fifth year, there is a minimal decrease in the indicator of the NDVI anomaly, relative to previous years. The difference between the fourth and fifth years is 2%. The total excess of background values over the values of disturbed areas in the fifth year was 31%, and the maximum values reached 40%.

During the vegetation season, the average background NDVI was at 0.6. Consequently, the decrease in the NDVI anomaly is caused by an increase in the NDVI indicators of the disturbed areas. The highest amplitude of NDVI values is observed in the first year of reforestation. Starting from the second year, the amplitude of the values is significantly reduced.

5 Conclusions

Extra insolation, as a consequence of reducing the albedo within the post-fire areas, induces extra heating of the underlying surface. It was instrumentally registered that the maximum excess of brightness temperature on post-fire areas can reach up to 11°C comparing to that of background areas under the same conditions. Such anomalies are determined by higher level of insolation due to partial or total tree mortality, as well as by decreasing of on-ground cover thickness after fire impact on grass and moss-lichen covers. During the vegetation season, averaged abnormal surface temperature is within the range of 4.74±1.0°C. The temperature maximum is decreasing by 29%, 42% and 51% during the second, fourth and the fifth year after fire respectively. However, the maximum temperature anomalies even after 5 years of vegetation recovery are staying at value of 4°C.

During the first year after a fire in larch forests of Siberia, the maximum temperature anomalies of the underlying surface was recorded in the third decade of June. In the course of 2–5 years after burning, the maximum temperature anomalies shift to the second or even third decade of July within the phenological season.

Analysis of NDVI indicators showed a trend of decreasing vegetation anomaly from the beginning to the end of the season in the first and second year of reforestation. During the fourth and fifth year of reforestation, the lowest anomalies were observed in the middle of the season (July). This means that the restoration of vegetation, starting from the fourth year, has reached a level at which the difference between background indicators and indicators of disturbed areas is reduced by increasing photosynthetic activity in disturbed areas. While during the first and second year of reforestation, this was due to a decrease in the photosynthetic activity of the background vegetation.

Based on the observed temperature anomalies and the NDVI anomalies could be concluded, that five years of post-fire succession is not enough to stabilize the disturbed plots relative to the background condition, both in the case of temperature changes and in the case of photosynthetic activity of vegetation. The suggested approach allows to assess the degree of fire impact on vegetation, as well as to predict changes in the active layer of permafrost soils, which may be a consequence of extra thermal flow at the surface in the circumstances of disturbed larch forests of Siberia.

Acknowledgments: This research was funded by the Russian Foundation for Basic Research (RFBR) and Government of the Krasnoyarsk krai, and Krasnoyarsk krai Foundation for Research and Development Support, grant number 18-41-242003 "Modeling and satellite monitoring of effects from thermal anomalies of the underlying surface in the seasonally thawed soil layer of the permafrost zone of Siberia".

References

1. Ponomarev E.I., Kharuk V.I., Ranson J. K. Forests, **7**, 1-9 (2016)
2. Shvidenko, A.Z. Schepaschenko, D.G., Contemp. Probl. Ecol., **7**, 50-61 (2013)
3. Valendik E.N., Kisilyakhov E.K., Ryzhkova V.A., Ponomarev E.I., and Danilova I.V. Geography and Natural Resources, **35**, 41-47 (2014)
4. Ponomarev E.I., Kharuk V.I Contemporary Problems of Ecology, **9**, 29-36 (2016)
5. Kharuk V.I., Ponomarev E.I. Russian Journal of Ecology, **48**, 507-512 (2017)
6. de Groot, W.J., Cantin, A.S., Flannigan, M.D., et al., For. Ecol. Manage., **294**, 23-34 (2013)
7. Flannigan, M., Cantin, A.S., and de Groot, W.J., For. Ecol. Manage., **294**, 54-61 (2013)
8. Ponomarev E.I., Ponomareva T.V. Contemporary Problems of Ecology, **11**, 420-427, (2018)
9. Tsvetkov P.A. Conifers of the boreal zone, **5**, 10-14 (2013)
10. Bezkorovaynaya, I.N.; Borisova, I.V.; Klimchenko, A.V.; Shabalina, O.M.; Zakharchenko, L.P.; Il'in, A.A.; Beskrovny, A.K. Vestn. KrasGAU, **9**, 181-189 (2017)
11. Brown, D.R.N.; Jorgenson, M.T.; Kielland, K.; Verbyla, D.L.; Prakash, A.; Koch, J.C. Remote Sens, **8**, 654 (2016)
12. Knorre, A.A.; Kirdyanov, A.V.; Prokushkin, A.S.; Krusic, P.J.; Buntgen, U. Sci. Total Environ, **652**, 314-319 (2019)
13. Bartalev, S.A., Stytsenko, F.V., Egorov, V.A., and Lupyan, E.A., Lesovedenie, **2**, 83-94 (2015)
14. Bondur, V.G., Gordo, K.A., and Kladov, V.L., Izvestiya, Atmos. Ocean. Phys., **53**, 859-874 (2017)
15. Veraverbeke S., Verstraeten W.W., Lhermitte S., Van De Kerchove R., Goossens R. International Journal of Wildland Fire, **21**, 243-256 (2012)
16. Cuevas-Gonzalez M., Gerard F., Balzter H., Rianos D. Global Change Biology, **15**, 561-577 (2009)
17. Krasnoshchekov K.V., Dergunov A.V., Ponomarev E.I. Sovremennye problemy distantsionnogo zondirovaniya Zemli iz kosmosa, **16**, 87-97 (2019)

18. Yakimov N.D., Ponomarev O.I., Ponomarev E.I. Actual problems of fire safety and emergency protection, 421-429 (2019)
19. Ponomarev E.I., Ponomareva T.V. Sovremennye Problemy Distantsionnogo Zondirovaniya Zemli iz Kosmosa, **15**, 85-95 (2018)
20. Ponomarev E., Ponomareva T., Masyagina O., Shvetsov E., Ponomarev O., Krasnoshchekov K., Dergunov A. Proceedings, **18**, 6-7 (2019)
21. Tsvetkov P.A. Conifers of the boreal zone, **23**, 186-195 (2006)

DYNAMICS OF CHANGES IN LONG-TERM AVERAGE ANNUAL RIVER RUNOFF IN THE LAKE SEVAN BASIN

Varduhi Margaryan[1,*], *Gennady* Tsibul'skii[2], *Ksenia* Raevich[2]
[1] Yerevan State University, Department of Physical Geography and Hydrometeorology, Faculty of Geography and Geology, 0025, Alek Manoukian 1 st., Yerevan, Armenia
[2] Siberian Federal University, Institute of Space and Information Technologies, Chair of Systems of Artificial Intelligence, 660074 Kirenskogo St. 26, Krasnoyarsk, Russia

Abstract. In the work, long-term fluctuations in the annual runoff of rivers flowing into lake Sevan were identified, analyzed and evaluated. As the source material, daily data of long-term actual instrumental observations of the annual runoff of rivers flowing into the lake, as well as air temperature and atmospheric precipitation of the lake basin for the entire observation period were used.

1 Introduction

Climate change, which began in the last century, will undoubtedly have its effect on the river runoff in Armenia. And lake Sevan is a strategic storage of fresh water of the Republic of Armenia, at the same time, the lake basin is one of the richest and most unique regions of the Republic of Armenia in terms of recreational resources.

The aim of the work was to investigate, analyze and evaluate the current multi-year fluctuations in the annual values of the average runoff of rivers flowing into lake Sevan, the average air temperature and the amount of atmospheric precipitation of the lake basin.

2 Study areas, Data and Methods

To solve the tasks posed, the corresponding scientific research works served as a theoretical basis. Actual data from Armhydromet's many-year observations of air temperature, precipitation and water discharge (from the data of odservation to 2018) were used as starting material. In the work, long-term fluctuations in air temperature and precipitation were discussed separately for meteorological stations, and changes in water flow rate for river sections of rivers. Note that over time, both the number of observed rivers and the number of hydrological posts, meteorological stations and posts located on them have changed.

28 rivers and 2 large springs flow into lake Sevan, 24 of them were carried out over the years in different years. Most of the rivers in the basin have a small length and catchment

* Corresponding author: vmargaryan@ysu.am

area. In the basin there are 930 rivers up to 10 km long, of which only Argichi is 51 km long. The relatively large rivers, whose catchment area is more than 100 km2, are as follows: Masrik, Karchakhpyur, Vardenis, Argichi, Bakhtak, Gavaraget (table 1).

The change in the quantitative and qualitative characteristics of water objects is most strongly influenced by direct anthropogenic impact, as well as a change in climatic conditions especially manifested in recent decades [5, 8, 16]. Assessment of these changes in various regions of the world, including the territory of the Sevan basin, is the subject of a number of works by various authors [4, 6–8, 12–14, 16].

Table 1. Basic hydrometric characteristics of rivers and their catchments in lake Sevan basin

River – point	Distance from the mouth, km	River length, m	Main catchment characteristics		
			Square, km^2	Average height, m	Average slope, ‰
Dzknaget – Tsovagyugh	1.0	22.0	82.6	2220	211
Masrik – Tsovak	2.8	45.0	673	2310	158
Karchaghbyur – Karchaghbyur	1.1	26.0	116	2650	174
Wardenis – Wardenik	4.1	28.0	117	2680	279
Martuni – Geghovit	7.2	27.0	84.5	2760	285
Argichi – Getashen	6.0	51.0	366	2470	144
Bakhtak – Tsakqar	6.0	30.0	144	2570	123
Gavaraget – Noratus	7.0	24.0	467	2430	133

In work mathematical-statistical, extrapolation, analytical, correlation methods are used.

3 Results and discussion

As a rule, the water regime of rivers is determined by the geographical position of the basin, relief, average height of the basin, geological structure, soil and vegetation cover, and nutritional conditions. Due to the influence of these factors, an uneven distribution of both the river network of the lake Sevan basin and runoff is manifested. The rivers of the northeastern part of the basin are significantly inferior in their water content to the rivers of the southern and western parts (Masrik, Argichi, Gavaraget) (table 2).

Table2. Main characteristics of the annual runoff of rivers flowing into lake Sevan

River – point	Average annual water consumption, m^3/sec	Drain module, l/sec.km^2	Drain layer height, mm	Maximum flow, m^3/sec	Minimum flow, m^3/sec
Dzknaget – Tsovagyugh	1.09	13.2	416	2.14	0.41
Masrik – Tsovak	3.33	4.95	156	4.38	2.15
Karchaghbyur – Karchaghbyur	1.06	9.14	287	1.52	0.37
Wardenis – Wardenik	1.58	13.5	425	3.17	0.47
Martuni – Geghovit	1.68	19.9	628	3.20	0.82
Argichi – Getashen	5.46	14.9	471	11.2	2.81
Bakhtak – Tsakqar	0.65	4.51	142	1.30	0.11
Gavaraget – Noratus	3.50	7.49	236	4.82	2.77

The work considers the time course and trends of the annual river runoff of the lake basin for the available observation periods for the year with an assessment of their significance (tig. 3, fig. 1), that is conclusions are drawn about the main trend of observed changes in river runoff.

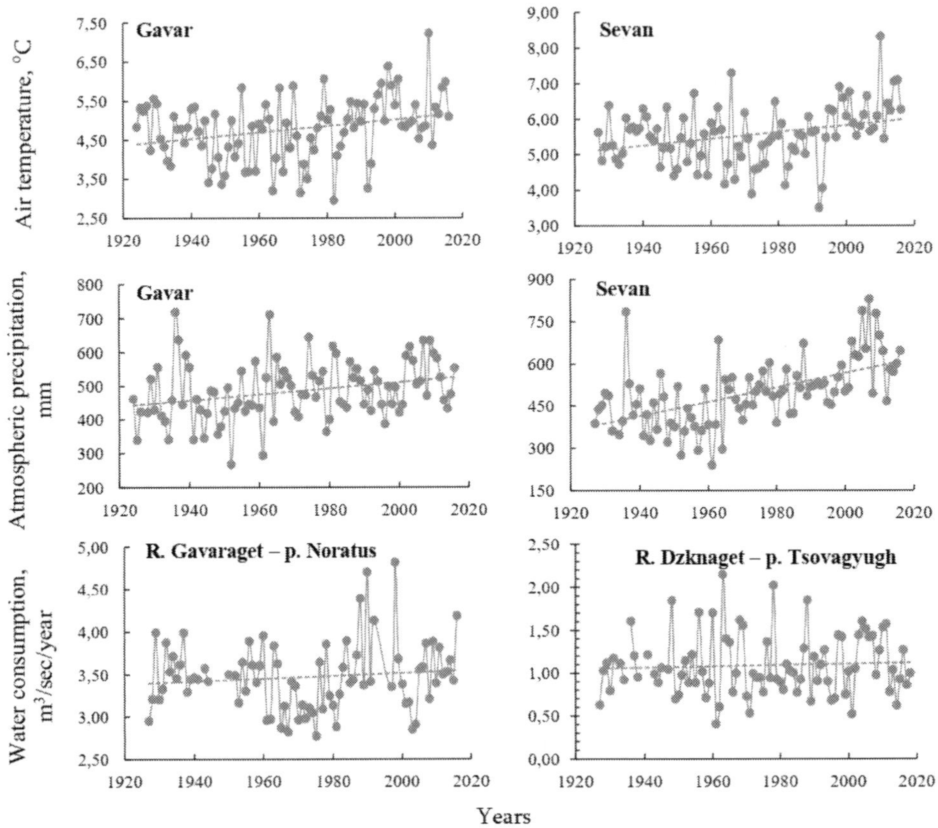

Fig. 1. Long-term fluctuations in average annual river water discharge, average annual air temperature, annual rainfall

Table 3 shows the equations of linear trends, statistical characteristics (trend coefficients), standard errors, etc). As follows from tab.3, the linear trend coefficient is the largest in the Masrik river – Tsovak point, which is equal $\beta = 0.11$ m3/sec / 10 years, which in terms of 100 years gives an increase in annual river runoff by 1.1 m3/sec.

For clarity, fig. 2 shows the graphs of the time course of the annual river runoff, on which trend lines are plotted. In the basin, according to the data of actual observations of the annual river runoff, they mainly show an upward trend. It turned out that in a number of long-term observations in the basin, a predominant growth trend was observed in annual river runoff. The only exceptions are the Vardenis and Bakhtak rivers, that is, a decreasing trend is observed by 17% of the rivers under discussion.

To assess the long-term fluctuations of the annual river runoff under the conditions of climate change, we also discussed the dynamics of changes in air temperature and precipitation. Changes in air temperature and precipitation are estimated for each meteorological station separately for the entire observation period.

Table 3. Linear trend equations and statistical characteristics for the annual river runoff of the lake Sevan basin

River – point	Linear trend equation	Statistical characteristics			
		β(m³/sek /10 years)	ΔQ_{90}, m³/sek	Autocorrelation, r(1)	Standard errors
Dzknaget – Tsovagyugh	y = 0.0007x + 1.053	0.007	0.063	0.02	0.039
Masrik – Tsovak	y = 0.0111x + 3.037	0.11	0.99	0.46	0.11
Karchaghbyur – Karchaghbyur	y = 0.0084x + 0.774	0.084	0.76	0.72	0.077
Wardenis – Wardenik	y = –0.0078x + 1.816	–0.078	–0.70	0.44	0.11
Martuni – Geghovit	y = 0.0037x + 1.562	0.037	0.33	0.44	0.098
Argichi – Getashen	y = 0.0078x + 5.093	0.078	0.70	0.25	0.19
Bakhtak – Tsakqar	y = –0.0005x + 0.663	–0.005	–0.045	0.02	0.031
Gavaraget – Noratus	y = 0.0008x + 3.467	0.008	0.072	0.20	0.060

The research results show that under the conditions of climate change at all meteorological stations of the basin territory, there is only a tendency for an increase in average annual air temperature (and significantly over the past decades) and annual precipitation. Over the entire observation period, the highest average annual air temperatures were observed in 2010 (fig.1). Long-term fluctuations in air temperature and precipitation also have their manifestation in fluctuations in the river runoff of the study area.

It turned out that in the study area over the entire period of instrumental observations the average annual air temperature increased by 0.5-1.5 °C or 10–3 0%, and atmospheric precipitation by 80-150 mm or 15–40 %.

Note that the study area has a mixed continental climate with a pronounced ascending zonality. Climate formation is greatly influenced by the isolation of the basin and the presence of the lake. As a result, being bordered by mountains, the amount of precipitation in the basin is relatively small. Winter is long, cold, stable snow cover forms. The summer is warm, the weather is mostly cloudy. In January, the average monthly temperature ranges from –4.5 °C (Shorzha) to –8.5 °C (Vardenyats), the absolute minimum reaches –38.1°C (Masrik). Summers are cool and short, in July the average monthly temperature is from 14.0 °C to 17.1 °C (Shorzha), the absolute maximum reaches 35.2 °C (Masrik).

In the lake basin, precipitation is distributed extremely unevenly [2]. In the northern and northwestern parts of the basin , a significant amount of precipitation falls than in other parts. This is explained by the fact that the northern and especially northwestern part is more accessible for the flow of external air masses. Annual precipitation in the mirror and coastal zones is 390–450 mm. Higher in the mountains, the amount of precipitation increases, reaching 800–900 mm. Most precipitation falls in the Dzknaget basin (635 mm – Tsovagyugh, 724 mm – Semenovka), and in the Masrik valley – relatively less (435 mm – Masrik).

4. Conclusions

Thus, under the current climate change in the study area, the average annual air temperature, annual precipitation, and average annual river discharge have increased over the entire observation period. In general, an increase in the amount of precipitation and river runoff in the study area in the development of various spheres and branches of the economy is assessed positively. Nevertheless, it is necessary to develop economic mechanisms for the efficient consumption and conservation of water resources, to create a full-fledged database of reliable data on water resources and climatic characteristics.

In the study area over the entire period of instrumental observations, the average annual air temperature increased by 0.5–1.5 °C or 10–30 %, and atmospheric precipitation by 80-150 mm or 15–40 %.

The reported study was funded by Russian Foundation for Basic Research (No.18-47-242002), Government of Krasnoyarsk Territory, Krasnoyarsk Regional Fund of Science, to the research project: «The development of technology for creating intelligent information systems of object-oriented monitoring of territories based on remote sensing data».

References

1. O.A. Azernikova Water balance of Lake Sevan according to new data. Proceedings of the All-Union Geographical Society. 1975. Volume 107, Issue 5, pp. 411
2. E. Alexanian The river flow cycle of the lake Sevan area // Geography 99, collected articles, Yerevan, 1999, pp. 77-84.
3. Andreyanov V.G. Cyclic fluctuations in annual runoff and their accounting in hydrological calculations. // Proceedings of the State Hydrological Institute. «Issues of runoff calculations». 1959. Issue 68. Leningrad: Hydrometeoizdat. P. 3–50.
4. Apukhtin A.V., Kumani M.V. Longstandig dynamics of basic elements of the spring drain of the small and middle rivers of the Central Chernozem region. Scientific bulletins of the Belgorod State University. Series: Natural Sciences. 2015, 21 (218): 114–120.
5. Vardanian T.G. Features of the changes annual flow of rivers Armenia. Geography and Natural Resources. 2007, 2: 169–177.
6. L.R. Vardanyan, K.H. Hayrapetyan, L.V. Azizyan, E.P. Eroyan, G.A. Hovasapyan, S.M. Mkrtchyan Assessment of seasonal and annual changes in precipitation and river flow in Sevan lake basin due to climate change (Republic of Armenia). Proceedings of the NAS RA, Earth Sciences. 2016, 69 (1): 38–49.
7. A.G. Georgiadi, Ye.A. Kashutina Regional features of the time changes in river runoff to the Arctic Ocean. Polar Cryosphere and Continental Waters. Moscow: Paulsen. 2011. P. 252–264.
8. A.G. Georgiadi, N.I. Koronkevich, I.P. Milyukova, A.V. Kislov, O.A. Anisimov, E.A. Barabanova, E.A. Kashutina, O.O. Borodin Scenario-based assessment of likely changes of river runoff in the basins of the largest rivers of Russia. Part 1. The Lena River basin. M.: 2011. 180 p.
9. N.V. Goroshko Methods of assessment of spatio-temporal fluctuations (based on the example of the upper Ob basin) // "Environmental dynamics and global climate change", Том 1 (2010), №1 (1).
10. V.K. Davydov The water balance of the lake Sevan. In the book: Materials on the study of Lake Sevan and its basin. L., part 6, 1938, 83 p.
11. V.G. Margaryan, T.G. Vardanian River drainage forming and intra-annual distribution features of Sevan Lake basin // Scientific and practical potential of geography and its practical application. Yerevan, YSU, 2011, Pp. 211-223.
12. V.O. Sargsyan Surface inflow in the lake Sevan. Collection of works of the Hydrometeorology Center ArmUGKS, Moskow, Hydrometeoizdat. Issue 1 (5), 1984, pp. 9-17.
13. N.L. Frolova, P.A. Belyakova, V.Yu. Grigoryev, A.A. Sazonov, L.V. Zotov Long-term fluctuations in river flow in the Selenga basin. Water resources. 2017, 44 (3): 1–13.
14. *Ecology of Lake Sevan during the period of water level rise the results of russian-armenian biological expedition for hydroecological survey of Lake Sevan (Armenia)* (2005–2009). Makhachkala: Science DSC, 2010. – 348 p.
15. M.V. Shaginyan, B.G. Zakaryan The share of river runoff in the water balance of Lake Sevan and expected changes due to climate change // Armenia. problems of climate

change / Collection of articles. Yerevan, "Science" Publishing House of NAS RA, 1999, p. 137-142.
16. N.M. Yumina, M.A. Tereshina Estimating long-term changes of river runoff in the Viluy river basin. Moscow University Bulletin. Series 5. Geography. 2017, (6):62-70.

On the issue of fluctuations in the extreme maximum runoff under the conditions of the expected climate change in the Marmarik river basin

Varduhi Margaryan [1,*], and *Elena* Fedotova[2,3]

[1]Yerevan State University, Department of Physical Geography and Hydrometeorology, Faculty of Geography and Geology, Yerevan, Armenia
[2]Sukachev Institute of forest SB RAS, FRC KSC SB RAS, Krasnoyarsk, Russian Federation
[3]SFU, Krasnoyarsk, Russian Federation

Abstract. The paper analyzes the peculiarities of formation of the absolute maximum runoff of the Marmarik river evaluates the patterns of multi-year fluctuations of maximum runoff rates in different river sites and gives a forecast of the maximum runoff in the context of global climate change. Absolute values of the maximum river runoff for different scenarios of climate change are estimated. The actual observational data of Armhydromet for maximum runoff rate, the air temperature and precipitation were used as the source material. As a result of the study, it turned out that there is only a tendency to decrease in the values of maximum runoff. It turned out that for all scenarios and cases in the Marmarik river basin, a different degree of changes in the maximum flow is observed. Moreover, the largest decrease in the maximum runoff of the Marmarik river basin is expected under the conditions of an increase in the average air temperature of the spring season by 2,7–3,9 degrees Celsius and a decrease in the amount of spring atmospheric precipitation by 2,4–2,6 %.

1 Introduction

The analysis and estimation of the multi-year fluctuations of extreme maximum runoff rates in the context of global climate change becomes of an increasing importance. Notably, it is essential for engineering purposes and the exploitation of the hydraulic constructions, as well as for the organization of damage-preventing measures. Since the manifestations of the current climate change may be observed both on global and regional scales, their evaluation is one of the primary tasks that concern humanity. Moreover, the analysis of the maximum runoff rates is of great importance as a guarantee for stable development and promising economic growth. It is necessary to highlight that the Marmarik river is the main and most water-abundant inflow of the Hrazdan river. The regulation of the Marmarik river water level permits to decrease the water releases from lake Sevan, which is a strategic

* Corresponding author: vmargaryan@ysu.am

fresh water storage site of the Republic. The river waters are exploited for irrigational purposes, as well as for supplying water to industrial enterprises. The river valley holds a large recreational potential, hosting a number of operating summer holiday houses. The Marmarik river valley contains a source of the mineral water "Hanqavan". It is the site of the yearly students' educational practice, such as hydrology.

Fig. 1. Marmarik river

Therefore, the aim of this investigation is to analyze and evaluate the patterns of fluctuation of the maximum runoff rates in the Marmarik river basin within the scope of the expected climate change. To reach this aim, the following research tasks have been formulated and solved:

– to process and evaluate the hydro- and meteorological elements of the basin;
– to determine and analyze the specificities of space-time distribution of the extreme maximum runoff rates of the basin rivers;
– to forecast and evaluate the maximum runoff in the context of the forecasted future climate change.

2 Study areas, Data and Methods

In order to complete the research tasks, numerous thematic literature on the work has been studied [1, 6, 8–12]. As a reference material, the research makes use of the observational results of the Service of the Hydrometeorology and Active Influence on Atmospheric Phenomena, Ministry of Emergency Situations of the Republic Armenia.

Hydro- and meteorological research has been conducted in the Marmarik river basin from the 1930s. Nevertheless, from the point of view of meteorological observations, it is one of the least developed territories. The unique meteorological station Hanqavan, located at an altitude of 1957 m, was closed in 2018, thus marking the end of continuous observations in the region. Therefore, the research utilizes observational data from the nearby meteorological stations – Hrazdan and Aparan. Regarding the hydrological observations, at present (year 2019), the basin contains solely three operating posts, the data from which was used in present research.

Notoriously, it is the physic-geographical conditions that determine the peculiarities of formation, sub-annual distribution and temporal change in the runoff, as well as the water and thermic regime. The above-mentioned conditions include: geological and hydrogeological structure, morphometric and morphological elements of the terrain, the average height of the basin, the water supply in the snow, the intensity of the snow-melt rate, the liquid precipitation and its infiltration in the soil, the duration of precipitation, its uneven distribution on the basin area and in time, the evaporation, the lakeiness, soil and vegetation cover, etc. The basin's terrain is of a mountainous type, largely divided by valleys, ravines and watercourses. The hydrogeological conditions are not favorable for the infiltration of surface water, and large underground water storages are absent. The basin climate is continental, with cold winter seasons and moderately warm summer.

The following methods were used in the present research: the mathematical statistical analysis, the compare and contrast method, extrapolation and correlation, as well as one of the most widely used methods – the normalized integral-difference curves.

3 Results and discussion

During the spring flood season, the extreme maximum runoff often takes place in the basin rivers. At this time, the runoff volume reaches 65–80% that of the whole year. The maximum water spending usually occurs in the end of April – beginning of May, especially the first half of May. The quantitative characteristics of the maximum basin river runoff are characterized by high variability. For instance, the extreme maximum water runoff in the Marmarik river – p. Aghavnadzor demonstrates significant variation between 9,85 m^3/sec to 86,7 m^3/sec. In order to reveal the key tendencies of the maximum runoff rate changes, the research used the methods of integral-difference curves, modelling and moving average by n-years. For all three operating hydrological points of the Marmarik basin rivers, the normalized integral-difference curves of the maximum runoff have been built (fig. 2). It is important to underline that throughout the last two decades the maximum runoff demonstrates a tendency for decrease in values.

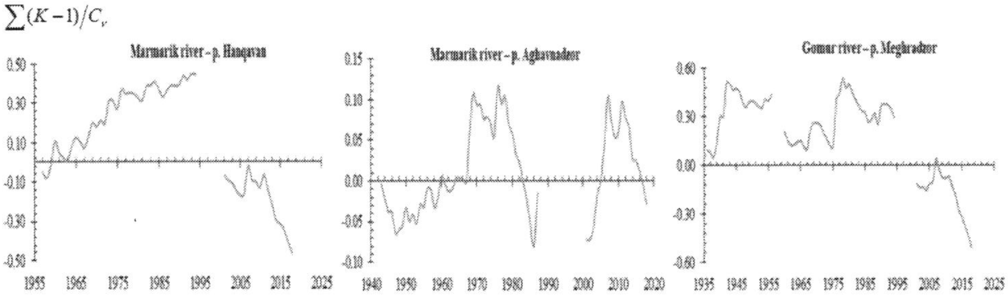

Fig. 2. The integral-difference curves of the extreme maximum water runoff in the Marmarik basin rivers.

As pointed out in the work [7], the normalized integral-difference curves are an effective tool for demonstration, allowing objectively identify the long-period cyclical fluctuations and their parameters. This is highly important for the choice of a representative period for the estimation of key parameters of the probability distribution of the maximum runoff. On fig. 3 it may be seen that throughout the observation period the change in the runoff was followed by a periodical change in the river water content. The change in the maximum runoff of the Marmarik river basin is also characterized by a certain periodic change in the water content of the rivers – cyclicity (low water-level and high water-level runoff phases, of short and long duration). One full cycle, where high and low water-level phases are distinguished, typically lasts 35–45 years containing 5–10 sub cycles. The synchronicity of the maximum runoff is weakly manifested. Fig.3 shows the normalized deviations of the maximum runoff in the p. Hanqavan of the Marmarik river. It may be clearly seen from the figure, that the decrease in the absolute values of the maximum runoff is followed by a decrease of the relation of the maximum runoff and the average multiple-year runoff.

In order to analyse the fluctuation tendencies of the maximum water runoff, the research additionally builds upon the linear trend method. The results demonstrate that a decrease in the maximum runoff may be observed on the researched territory (a negative tendency). In other words, it may be assumed that the forecasted hydrological risk level from natural

 https://doi.org/10.1051/e3sconf/202014903010

disasters rapidly diminished. Similar changes are typical for several other rivers on the CIS territories [3–5, 7, 10–11, 13, 15].

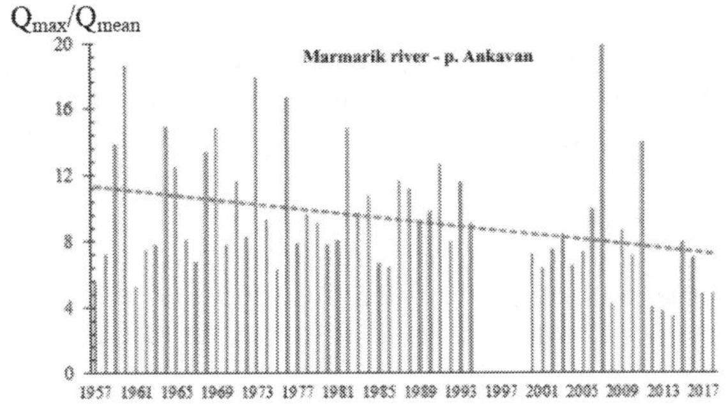

Fig. 3. The normalized deviation of the Marmarik river p. Hanqavan maximum runoff values, throughout the observation period.

This circumstance is mainly due to an increase in air temperature in the cold season (fig. 4), and as a result, an increase in the frequency of thaws, diminishing heights of the snow cover and water storage in the snow layer at the beginning of snow-melt period. The diminishing norm of the absolute value of maximum runoff in the investigated rivers throughout the observational period fluctuates largely, from 18% to 50% (table 1).

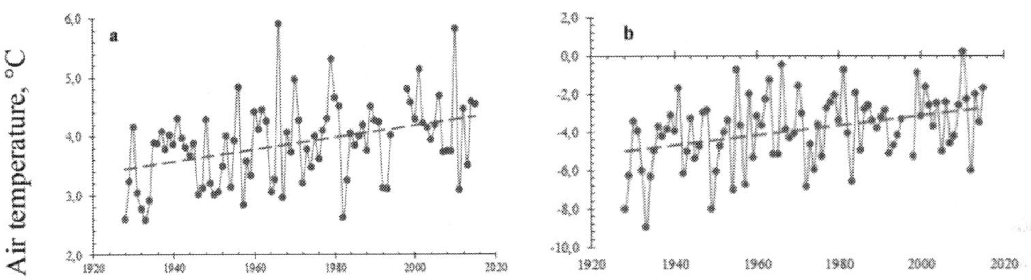

Fig. 4. Inter-year fluctuations of the average air temperature values during the cold season period, built from data of the Hanqavan (a) and Hrazdan (b) weather stations.

Table 1. Change in the characteristics of the absolute maximum runoff values.

River – hydrological point	Linear trend equation	Correlati-on coeffi-cient, r^2	Norma of the maximum runoff, m^3/sec	Runoff changes	
				m^3/sec	%
Marmarik – Hanqavan	y=–0,112x+238,51	0,107	15,7	– 6.38	– 40,7
Marmarik – Aghavnadzor	y=–0,098x+232,03	0,014	39,1	– 6,96	– 17,8
Gomur – Meghradzor	y=–0,096x+204,25	0,082	14,9	– 7,30	– 49,0

For the purposes of estimating the influence of the climate change on the maximum runoff, regression models have been used. They describe the linear dependencies between the runoff elements and climatic factors (such as the air temperature and atmospheric precipitation levels).

To determine the preliminary or climatic basis state, to be used for comparison with the future forecasts, a large dataset is required, which would include the climate volatility cycles and which would characterize the modern climatic conditions. For an estimation or forecast of the influence of climate change, a quantitative assessment of the future climate change is a prerequisite. At present, there is a lack of methods allowing to perform robustly such assessment. Instead, a so-called set of the future climate scenarios is used. According to the CCSM4 model, the scenarios of the RCP8.5 (equivalent to the SRES A2 scenario) and RCP6.0 (equivalent to SRES B2) emissions for Armenia region, the future changes in air temperature and precipitation levels until year 2100 were developed [2]. Table 2 shows the forecasted changes of the average air temperature and precipitation levels on Armenian territory for the spring period, compared to the average for 1961-1990 period. From the research side, it was taken as a basis of estimation for the Marmarik river maximum runoff reaction in relation to the forecasted climate change.

Table 2. The forecasted change in the average air temperature and the summative atmospheric precipitation levels on the territory of Armenia, throughout the spring period, in relation to the average for 1961–1990 period.

Scenarios	2011–2040	2041–2070	2071–2100
Average air temperature, °C			
RCP6.0	1,3	2,4	2,7
RCP8.5	1,4	2,7	3,9
Precipitation, %			
RCP6.0	1,2	4,2	2,6
RCP8.5	4,2	- 8,0	2,4

Note that, according to the CCSM4 model, the results of the future climate change in general correspond to the previously obtained results on the regional model PRECIS. Nevertheless, the CCSM4 model results are remarkably more successful for temperature than for precipitation levels.

To evaluate the runoff reaction, the relation between the runoff values and the climatic factors (air temperature and atmospheric precipitation levels throughout March-May period) has been examined. As a following step, the changes of the absolute maximum runoff norm in comparison to the average runoff values have been evaluated for the 1961–1990 period (table 3). The estimation was performed based on factual data and climatic scenarios for 2011–2040, 204–2070 and 2071–2100 time intervals. The table demonstrates that before the end of the current century, a clear tendency for decrease may be noted in the year-on-year changes of the maximum runoff norm. Most likely, this is due to the increase in the air temperatures. Such difference of the maximum runoff regime is marked as typical, as well, in the Arctic region [14], and several other rivers on the CIS territories.

Within various scenarios, a different magnitude of changes in the maximum runoff may be observed in the Marmarik river basin. The minimal changes were observed in the first scenario, maximal – in the second. The maximal changes in the average values of the extreme runoff levels, in comparison to the average values for 1961–1990 period, were observed in the Aghavnadzor point of the Marmarik river and Meghradzor of the Gomur river. The corresponding value changes were 13–52% and 19–59%, accordingly.

In conclusion, the extreme maximum runoff in the Marmarik river basin often occurs during the spring flood season. For all investigated rivers of the basin during the whole observational period, a directed decrease of 18% and higher of the maximum runoff norm (due to climate change) is a typical tendency. Consequently, the danger of flooding has rapidly diminished. The variance of the maximum runoff in the basin rivers is characterized by cyclicity. A full cycle corresponds to a 35–45 year period, containing 5–10 sub cycles.

Table 3. The forecasted changes in the average values of the Marmarik river maximum runoff, in comparison to the average values throughout 1961-1990 period.

River – hydrological point	Scenarios	Runoff changes							
		1961–1990		2011–2040		2041–2070		2071–2100	
		m³/sec	%	m³/sec	%	m³/sec	%	m³/sec	%
Marmarik – Hanqavan	RCP, 6.0	16,9	0	–2,80	–16,6	–5,2	–30,8	–6,00	–35,5
	RCP, 8.5	16,9	0	–3,00	–17,8	–6,2	–36,7	–8,70	–51,5
Marmarik – Aghavnadzor	RCP, 6.0	37,9	0	–5,01	–13,2	–10,6	–28,0	–12,7	–33,5
	RCP, 8.5	37,9	0	–4,81	–12,7	–15,6	–41,2	–19,7	–52,0
Gomur – Meghradzor	RCP, 6.0	15,6	0	–3,00	–19,2	–5,30	–34,0	–6,30	–40,4
	RCP, 8.5	15,6	0	–2,90	–18,6	–7,60	–48,7	–9,20	–59,0

The reported study was funded by Russian Foundation for Basic Research (No. 18-47-242002), Government of Krasnoyarsk Territory, Krasnoyarsk Regional Fund of Science, to the research project: "The development of technology for creating intelligent information systems of object-oriented monitoring of territories based on remote sensing data".

References

1. A.V. Apukhtin, M.V. Kumani, Scientific bulletins of the Belgorod State University. Series: Natural Sciences. **21 (218),** 7 (2015)
2. *Armenia's Third National Communication on Climate Change.* Ministry of Nature Protection of the Republic of Armenia. (Yerevan: Lusabats. 2015)
3. N.L. Frolova, P.A. Belyakova, V.Y. Grigoryev, A.A. Sazonov, L.V. Zotov, Water resources. **44 (3),** 13 (2017)
4. N.L. Frolova, M.B. Kireeva, S.A. Agafonova, V.M. Evstigneev, N.A. Efremova, E.S. Povalishnikova, Water sector of Russia **4**, 17 (2015)
5. A.M. Gareev, P.N. Zaytsev, Herald of the Academy of Sciences of the Republic of Bashkortostan. **21(1)**, 8 (2016)
6. *Hydrography of Armenian SSR.* Yerevan: Publishing House of the Arm. SSR Academy of sciences (1981)
7. F.A. Imanov, N.I. Hasanova, *Proceedings of the Russian State Hydrometeorological University.* **9**, 6 (2009)
8. *IPCC: Climate Change 2007: The Physical Science Basis.* Cambridge University Press, Cambridge, USA (2007)
9. D.V. Magritsky, *River runoff and hydrological calculations: practical work with the implementation of computer programs.* (M.: Triumph Publishing House, 2014)
10. V.G. Margaryan, Water Sector of Russia. **6** (2019)
11. V.G. Margaryan, Hidrolohiia, hidrokhimiia i hidroekolohiia: The scientific collection. **4 (51)**, 8 (2018)
12. *Surface-Water Resources of the USSR* **9**(2), (L.: Hydrometeoizdat, 1973)
13. G.R. Safina, V.N. Golosova, *Proceedings of Kazan University. Natural Sciences Series.* **160 (1)**, 14 (2018)
14. Shevnina E.V., Construction of Unique Buildings and Structures. **7 (22),** 13 (2014)
15. T.G. Vardanian, *Proceedings of the Russian State Hydrometeorological University.* **46,** 8 (2017)

Features of atmospheric disturbances in temperate latitudes before strong earthquakes (M> 7) according to satellite measurements

Valentin Kashkin[1], *Leonid* Sverdlik[2], *Roman* Odintsov[1], *Tatyana* Rubleva[1], *Konstantin* Simonov[3], *Alexey* Romanov[1] and *Sanjar* Imashev[2]

[1]Siberian Federal University, Krasnoyarsk, 660074, Russia
[2]Research Station of the Russian Academy of Sciences in Bishkek, Kyrgyzstan
[3]Institute of Computational Modeling SB RAS, Krasnoyarsk, Russia

Abstract. An ATOVS and MERRA data archive containing temperature values at isobaric levels from 1000 to 50 hPa was formed to study the atmospheric effects observed over the foci of two strong earthquakes with magnitudes M = 7,3. The first seismic event was recorded on August 19, 1992 in the northern Tien Shan and the second on November 12, 2017 in the border areas of the Zagros folded mountain belt. According to USGS intense aftershock activity in these regions began with these seismic events. Using correlation and spectral analysis we studied the vertical profiles of temperature and pressure in a disturbed atmosphere over seismically active regions. The modified anomaly index was calculated at isobaric levels from 500 to 50 hPa and on its basis the registered temperature anomalies were investigated.

1 Introduction

Using remote sensing data makes it possible to solve the problems of monitoring atmospheric phenomena observed in the geological environment during seismic processes. These are atmospheric effects over fault zones of the earth's crust in the form of perturbations of meteorological fields and temperature anomalies [1, 2]. In [1] it is indicated that the movements of the earth's crust excite internal gravitational waves in the atmosphere. These processes are accompanied by various variations of geophysical fields and contain valuable information about the features of the observed anomalous phenomena. In this regard, it becomes relevant to study the interaction of the lithosphere-atmosphere during geodynamic activity.

In our work according to the satellite equipment ATOVS (Advanced TIROS Operational Vertical Sounder, KA NOAA / POES, USA) and the global reanalysis database MERRA (Modern-Era Retrospective Analysis for Research and Applications, NASA), available on sites [3, 4]. The atmospheric temperature variations during the sharp activation of seismic processes are analyzed. At that time in the mountainous regions of the northern Tien Shan (Kyrgyzstan) on August 19, 1992 and in the north-west of the Zagros mountain belt (Iraq-Iran) on November 12, 2017 strong earthquakes with magnitude M = 7,3 were recorded.

2 Seismic data

In the mountainous region of the northern Tien Shan a strong earthquake of magnitude M = 7,3 occurred at 02:04:37 UTC on August 19, 1992 [5]. Note that the northern Tien Shan is a seismically active intracontinental mountain belt located in the zone of collision of the Eurasian and Indian lithospheric plates. However, the epicenter of the main shock with coordinates (42.142° N; 73.575° E) was in the seismically weakly active region of the Suusamyr depression. The focus of the hypocenter was located at a depth of about 27 km.

A fragment of the map of the aftershock field of the Suusamyr earthquake is shown in Fig. 1, *a*. After the seismic event M=7,3, three strong aftershocks (M≥6,1) occurred in the area within 50 minutes, which arose in the Earth's crust with boundaries of 18.5–33.0 km. In total, during the day, August 19, 1992, 20 aftershocks with magnitudes of 3,8<M<6,3 were recorded here. Only by the end of 1993 as follows from the data of [6] the seismicity in the study area gradually began to decrease to background values.

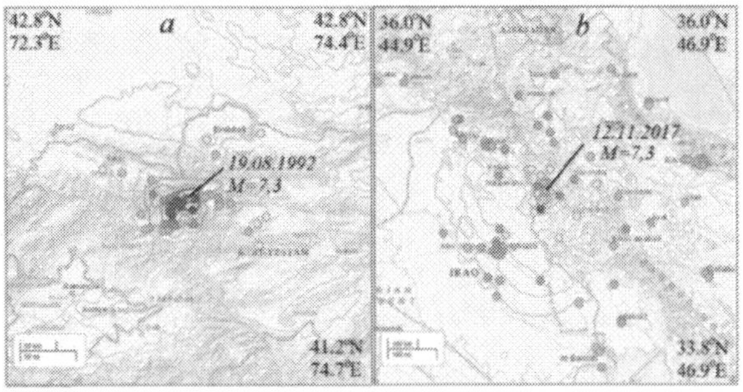

Fig. 1. Aftershock field of the earthquake August 19, 1992 in the northern Tien Shan (*a*) and location of the aftershock field of the seismic event November 12, 2017 in the Iraq-Iran (*b*)

The second seismic event of magnitude M = 7,3 (Fig. 1, *b*) occurred on November 12, 2017 at 18:18:17 UTC in the north-west of another intracontinental mountain belt Zagros located in the Middle East in the border regions of Iraq-Iran [7, 8] . As indicated in [7] the fold-thrust belt Zagros was formed as a result of a collision of the Arabian and Eurasian tectonic plates. Undoubtedly, the seismicity of the area where a strong earthquake M = 7,3 occurred is directly related to the geodynamic processes in the zone of collision of these two plates.The aftershock field of this strong earthquake is shown in Fig. 1, *b*. The epicenter of the main shock with coordinates 34.91° N and 45.96° E was located 29 km south of the city of Halabadjan (Iraq) [7]. The hypocenter of the earthquake M = 7, 3 was at a depth of ~19 km. According to the USGS we determined the boundaries of the peak region of the Iraq earthquake. Its dimensions in latitude were 33.8–36.0° N and in longitude 44.9–46.9° E.

Analysis of the seismic data [5] of the aftershock field of the Iraq earthquake showed that the geodynamic activity in the studied peak region began with the earthquake M = 4,1 recorded on November 3, 2017 at 10:33:41 UTC. Over the next 9 days no seismic events were recorded in this area, i.e. seismic lull was observed. Only on November 12, 2017 40 minutes before the main shock at 17:35:31 UTC a new foreshock of magnitude M = 4,6 was registered here. In total in November 2017 40 aftershocks occurred in the focus of this earthquake. The depths of their hypocenters varied between 7.0–19.0 km. The highest seismic activity in the peak region according to the USGS was observed in 2018. This year 73 earthquakes with magnitudes of 3,7<M<6,3 occurred here.

3 Satellite data and analysis technique

The study of atmospheric disturbances over the studied seismic regions was carried out according to the ATOVS and MERRA data [3, 4]. ATOVS satellite information is obtained from measurements with HIRS-4 AMSU-(A, B) and MHS instruments. Temperature profiles were restored using the IAPP (International ATOVS Processing Packages) software package by solving the inverse problem [3]. ATOVS data has a spatial resolution of 1°×1° in latitude and longitude. As a result we formed temperature series on isobaric surfaces from 1000 to 100 hPa with a time step of 3 hours. The standard deviation of ATOVS data in temperature from the results of measurements of radiosondes is 1.3 K.

The MERRA global meteorological reanalysis database is a modern global model of GMAO (Global Modeling and Assimilation Office) which is constantly being adjusted taking into account satellite measurements of the GEOS-5 (Goddard Earth Observing System, NASA) and the results of observations of a network of ground stations in order to ensure reliability forecast calculations [4]. MERRA data have a spatial resolution of 0.5°×0.675° (latitude-longitude).

The technique of processing temperature series was considered in detail in [2, 8]. It is based on the correlation and spectral analysis of temperature profiles in atmospheric layers from the underlying surface to the lower stratosphere. Using our methodology we studied the features of the amplitude, frequency and phase of temperature variations observed before strong earthquakes. To detect temperature anomalies over areas where earthquakes are recorded the RST (Robust Satellite Techniques) method is widely used [9]. We calculated the modified anomaly index δT for a perturbed atmosphere taking into account the measurement time and pressure at a certain isobaric level [2, 8]. Then the integral indicator of temperature anomalies $D_{\delta T}$ was calculated which was found from the relation:

$$D_{\delta T} = DISP_{\delta T}^{ut} \cdot DISP_{\delta T}^{tr},$$

where $DISP_{\delta T}^{ut}$ is the variance of the moving average series of the anomaly index δT in the upper troposphere (upper troposphere – *ut*) and $DISP_{\delta T}^{tr}$ is the variance of the moving average series of the δT index in the tropopause (tropopause – *tr*).

The values of the integral parameter $D_{\delta T}$ exceeding unity indicate the presence of antiphase anomalous in amplitude temperature variations in both considered atmospheric layers. Further $D_{\delta T}$ values were compared with data on seismic activity in the studied areas.

4 Temperature variations

The application of the above methodology allowed us to identify common signs and features of atmospheric effects observed during the preparation of strong earthquakes with M = 7,3 in the northern Tien Shan and the folded mountain zone Zagros. Based on the obtained temperature profiles at isobaric levels from 500–50 hPa atmospheric disturbances over areas of strong earthquakes were studied during the period of seismic processes in August 1992 and November 2017.

In Fig. 2, *a* shows the temperature profiles T_{300} at the isobaric level of 300 hPa (upper troposphere) and T_{100} at the level of 100 hPa (lower stratosphere) during the preparation of the earthquake at the main seismic moment, and during aftershock activity over the seismically active region of the northern Tien Shan restored according to the data reanalysis MERRA. It was found that approximately 5–8 days before seismic events, positive temperatures T_{100} and negative temperatures T_{300} are observed.

In Fig. 2, *b* shows graphs of the dependence of temperature on pressure for August 9, 14 and 18, 1992 when preparing the main shock M = 7,3. The most amplitude changes in temperature are found in the atmospheric layers of 300–250 hPa and 150–100 hPa.

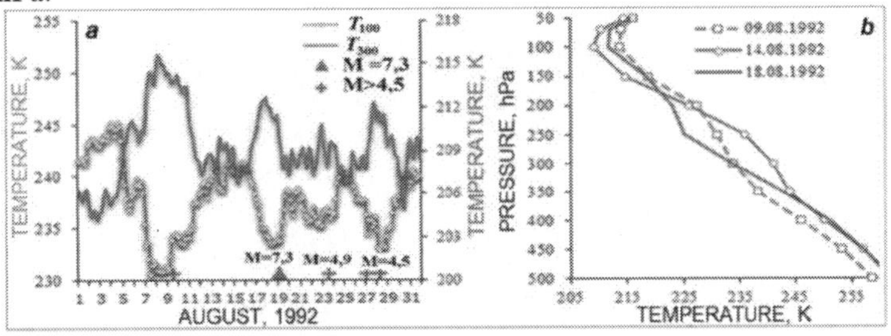

Fig. 2. Temperature series T_{300} and T_{100} in August 1992 (*a*) and vertical temperature profiles before the Suusamyr earthquake (*b*) according to MERRA

In Fig. 3, *a* shows the graphs of the temperature series T'_{300} (at the level 300 hPa) and T'_{100} (at the level 100 hPa) in a disturbed atmosphere over the border regions of Iraq-Iran during the November 2017 seismic processes, constructed using ATOVS satellite data. The correlation coefficient between the series T'_{300} and T'_{100} $R = -0.68$, consequently temperatures in the upper troposphere and lower stratosphere above the peak region of the Iraqi earthquake change in antiphase. The transition of the correlation function through "0" and a change in the sign of the correlation coefficient in the atmosphere disturbed by seismic events in the studied period of 2017 occurred near the tropopause in the transition zone of 300–200 hPa.

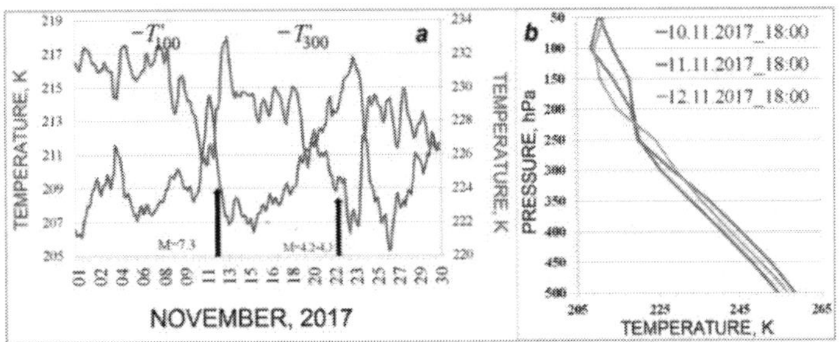

Fig. 3. Temperature series T'_{300} and T'_{100} (*a*) and temperature profiles before the Iraqi earthquake (*b*) in November 2017 according to ATOVS

If we consider temperature as a characteristic, then we can see the change in pressure over time in a disturbed atmosphere. In Fig. 3, *b* show examples of vertical temperature profiles on November 10, 11 and 12, 2017. It was found that the most significant temperature changes occur in the atmospheric layer of 300–100 hPa.

In Fig. 4, *a* shows the temperature series T_{300}, T_{200} and T_{100} for 1992 reconstructed according to the MERRA reanalysis, at the corresponding isobaric levels of 300, 200 and 100 hPa. It can be seen that from June to October 1992 antiphase temperature variations were observed.

In Fig. 4, *b* shows a graph of the correlation coefficient between the series dT_{300} and dT_{100}, characterizing the temperature increments $dT=T(t_{i+1}) - T(t_i)$ at isobaric levels of 300 and 100 hPa, respectively. Series dT_{300} and dT_{100} were calculated for the same time in a sliding window $m = 40$ days. Before a strong earthquake on August 19 in Fig. 4, *b* one can see a decrease in the correlation coefficient *R* to 0.84. This is probably due to two circumstances: either with an increase in the amplitude of temperature fluctuations or with the synchronism of antiphase temperature variations.

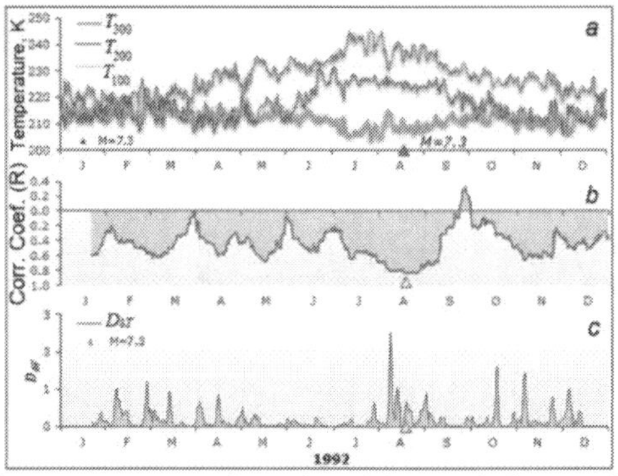

Fig. 4. Temperature series at 300, 200 and 100 hPa (*a*), correlation coefficient changes between series dT_{300} and dT_{100} (*b*) and anomaly index variations $D_{\delta T}$ (*c*) for 1992

In Fig. 4, *c* graph of the $D_{\delta T}$ anomaly index for 1992 is shown in the layers separated by the tropopause of 300–250 hPa and 150–100 hPa. Values $D_{\delta T}$ were calculated for the same moment in a sliding window m = 10 days. As can be seen in Fig. 4, *c* in 8 days before the Suusamyr earthquake the $D_{\delta T}$ index increased to 2.5.

5 Temperature anomalies

To detect anomalies observed over seismically active regions maps of the spatial distribution of daily average values of outgoing long-wave radiation (OLR) for August 1992 were constructed. Values δW $D_{\delta T}$ for each pixel were determined using the RST method [9] as the deviation of the current value of OLR from its five-year average values for 1987–1991, normalized on the standard deviation σ.

In Fig. 5 shows examples of anomalies in the outgoing long-wave radiation for August 15 and 18. It was found that the OLR anomaly in Fig. 5, *a* observed on August 15, 1992 and exceeding the 2σ level approximately repeated the location of active regional faults which are shown by dashed lines. This anomaly is elongated in the sub-latitudinal direction. On August 18, 1992 about a day before the Suusamyr earthquake the OLR anomaly was recorded near the epicenter of the main seismic shock (Fig. 5, *b*).

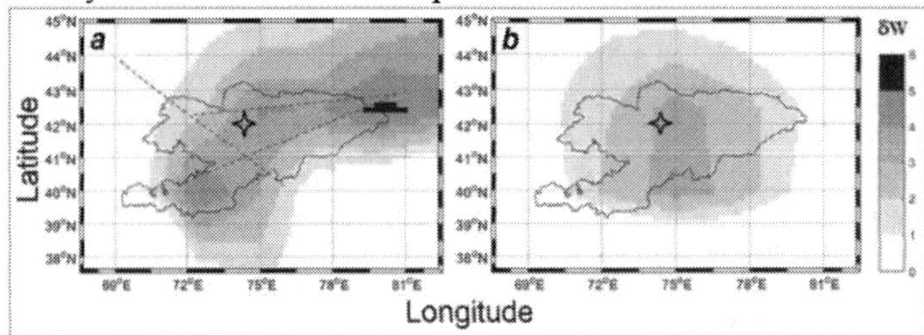

Fig. 5. OLR anomalies of near the epicenter of the earthquake on August 15 (*a*) and August 18 (*b*) 1992

In Fig. 6 shows examples of digital maps of the difference in temperature fields and the corresponding isotherms at 300 hPa and 100 hPa for November 12, 2017. The temperature field is shown at 25-45° N and 34-65° E. The temperature deviation scale ΔT is shown on the right. The epicenters of strong earthquakes are indicated by bold dots, their focal areas are indicated by a solid line, and the preparation zones are indicated by a dashed line. Here you can see the temperature anomalies existing at this moment.

In Fig. 6, *a* the temperature deviation scale is designated as ΔT_{1-2} where $\Delta T_{1-2}=T_1-T_2$. Here T_1 is the temperature in disturbed atmospheric conditions for November 12, 2017 at an isobaric level of 300 hPa, and T_2 is the temperature at the same level for the background mode for November 12, 2016. The anomalous region with negative temperatures ΔT_{1-2} is shown in Fig. 6 a. It is oriented in the latitudinal direction to the west-east. The epicenter is in the area of condensation of isotherms.

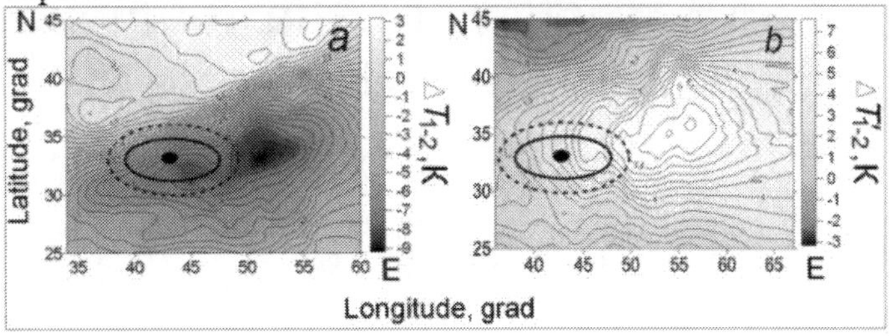

Fig. 6. Difference maps of the temperature field at the levels of 300 hPa (*a*) and 100 hPa (*b*) for November 12

In Fig. 6, *b* the temperature deviation scale is designated as $\Delta T'_{1-2}$ where $\Delta T'_{1-2}=T'_1-T'_2$. Here T'_1 is the temperature in disturbed atmospheric conditions for November 12, 2017 at an isobaric level of 100 hPa and T'_2 is the temperature at the same level for the background mode for November 12, 2016. Anomaly with positive values $\Delta T'_{1-2}$ is located in the lower stratosphere (Fig. 6, *b*). It is also oriented in the latitudinal direction to the west-east.

6 Conclusion

Using remote sensing data makes it possible to solve the problems of monitoring atmospheric phenomena observed in the geological environment during seismic processes. These are atmospheric effects in the form of disturbances in the parameters of meteorological fields and temperature anomalies over fault zones of the earth's crust. The results of the study of temperature fields in the upper troposphere and lower stratosphere indicate the presence of a relationship between the observed disturbances in the atmosphere of temperate latitudes and the activation of seismic processes in areas belonging to two different mountain systems. The presented results are a continuation of the study of the disturbed atmosphere during seismic processes according to the data.

References

1. V.B. Kashkin, Atmospheric and Oceanic Optics **26**, 10 (2013)
2. L.G. Sverdlik, S.A.Imashev, J. Sib. Fed. Univ. Eng. Technol.**10**, 6 (2017)
3. NOAA. URL: http://www.arl.noaa.gov/ready/cmet.html
4. NASA. URL: http://giovanni.gsfc.nasa.gov/
5. USGS. URL: https://earthquake.usgs.gov/earthquakes/map
6. Z.A. Kalmetieva, A.V. Mikolaichuk, and et al. Bishkek: CAIAG. (2009)
7. B. Mahani, A.J. Kazemian, *J.Seismol.* **22** (2018)
8. V.B. Kashkin, R.V. Odintsov, T.V. Rubleva, A.A. Romanov, K.V. Simonov, Trigger effects in Geosystems (Moscow, June 4-7, 2019): Materials of the V International Conference / M.:Torus Press (2019)
9. V. Tramutoli, V. Cuomo, C. Filizzola, N. Pergola, C. Pietrapertosa, Remote Sensing of Environment, **96,** 3-4 (2005)

Atmospheric reaction to the manifestations of the geodynamic activity of the Sulawes phenomenon according to satellite data

Valentin Kashkin[1], *Roman* Odintsov[1], *Tatyana* Rubleva[1 *], *Konstantin* Simonov[2], and *Julia* Tsup[1]

[1]Siberian Federal University, Krasnoyarsk, 660074, Russia
[2]Institute of Computational Modeling SB RAS, Krasnoyarsk, Russia

Abstract. Using ATOVS satellite data (NOAA / POES) atmospheric disturbances that arose in the equatorial zone of Indonesia in the fall of 2018 were studied as a reaction to the geophysical manifestations of the geodynamic activity of a strong earthquake with a magnitude of M = 7,5. An archive of satellite information has been formed. A technique has been developed for analyzing the temperature of profiles during strong seismic activity of the Sulawes phenomenon. The atmospheric effects over the seismically active region of this earthquake were studied. It was found that in the troposphere during the study period anomalies with lower temperatures are observed above the epicenter region of the Indonesian earthquake and thermal anomalies with elevated temperatures are formed in the lower stratosphere.

1 Introduction

The study is devoted to an actual problem – the study of the effects of the interaction between the lithosphere and the atmosphere during geodynamic activity. Satellite data make it possible to detect phenomena in a disturbed environment at different stages of geophysical intensification in the form of variations in temperature, pressure and humidity on various isobaric surfaces.

In this paper we consider the results of a study of atmospheric disturbances that arose in the fall of 2018 over the seismically active region of Sulawesi Island (Indonesia) based on remote sensing data ATOVS (Advanced TIROS Operational Vertical Sounder, NOAA / POES). In this geodynamically active place of the Asia-Pacific region in September-October 2018 three catastrophic geophysical events occurred consecutively during the week: an earthquake with M = 7,5, a tsunami in Palu Bay and an eruption of the Soputan volcano [1, 2]. The aim of the study is to obtain estimates of the geophysical manifestations of the Sulawes phenomenon.

* Corresponding author: tvrubleva@mail.ru

2 Analysis of seismic activity in the region

The seismically active region of Sulawesi Island is located in the south-western part of the Pacific subduction zone in the area of interaction between the Australian, Eurasian and Philippine tectonic plates [3, 4].

September 28, 2018 on about Sulawesi Island at a depth of 10-20 km seven seismic events with magnitudes of $5{,}8 \leq M \leq 7{,}5$ occurred. The strongest of them with M = 7,5 was recorded at 10:02:45 (UTC) with a focal depth of 20 km [1, 5]. The coordinates of the epicenter of this strong earthquake, which occurred approximately 80 km north of Palu, are 0.256°S, 119.846°E. The focal region was 200 km long and 20 km wide. The geodynamic features of seismic events M = 7,5 are considered in detail in [2, 6]. The likely cause of the Sulawes earthquake was the movement of the earth's crust along the Palu-Koro Fault.

According to IRIS [7] from September 21 to October 4, 2018 49 seismic events with magnitudes of $4{,}3 \leq M \leq 5{,}8$ were recorded in the aftershock field of the Sulawes earthquake. The distribution of magnitudes of foreshocks and aftershocks with the development of the seismic process in time is shown in Fig. 1. Most of the hypocenters of these foreshocks and aftershocks were located at a depth of 10-20 km. The strongest foreshock of magnitude M = 6,1 was recorded 3 hours before the main seismic event M = 7,5. As can be seen from Fig. 1 after the main shock in the focal region, the aftershock activity decreases, which is described by the Omori law [8].

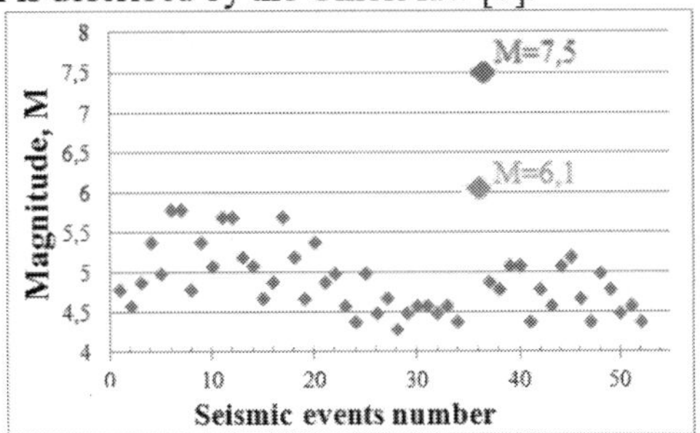

Fig. 1. Forshock and aftershock sequence of earthquake seismic events M = 7,5.

The chronology of geophysical events on September 28 shows that 7-12 minutes after the earthquake with M = 7,5 on the coast of about. Sulawesi hit the first tsunami waves [1, 2], two waves were recorded at intervals of 5 and 10 minutes. Along the coast of the island the wave height varied from 1.5 to 2 m. In the Palu Bay surrounded by mountains with peaks reaching up to 2000 m and which became a "funnel" for these waves, the tsunami splash height varied from 6 to 10 m.

After 4 days on October 2 at 22:46 (UTC) an increase in the amplitudes of seismic signals in the northeastern part of Sulawesi Island in the area of the Soputan stratovolcano with coordinates 1°108'N, 124°73'E [9]. His eruption occurred on October 3 at 08:44 UTC. Volcanic Explosivity Index [10] Soputan which describes the relative measure of explosiveness of volcanic eruptions and is used to estimate their impact on the atmosphere is 3. According to seismic information this geophysical event lasted 6 minutes [1]. As a result of the eruption, an ash plume formed rising to an altitude of 5.8 km (above sea level) and shifting towards the west and north-west.

The interconnectedness of the above natural disasters made it possible to introduce the term Sulawes phenomenon to characterize these geophysical events. As a result of the analysis of the development of seismic processes carried out according to the USGS and IRIS catalogs [5, 7] the period of active existence of the Sulawes phenomenon was determined – this is the time interval from September 21 to October 4.

3 Analysis of the reaction of the atmosphere to the developments of geodynamic activity

In our work we studied in detail the effects of the atmospheric reaction on the geophysical developments of the geodynamic activity of a strong earthquake with a magnitude of M = 7,5 in autumn 2018 obtained using the ATOVS instrument complex and presented on the site [11]. ATOVS equipment is designed to restore atmospheric meteorological fields at different heights. The ATOVS instrument complex measures the intensity of the Earth's own radiation in individual absorption bands of gases in the infrared and radio ranges [12]. The ATOVS hardware span is 2,250 km. The spatial resolution is 20-50 km horizontally and 2-3 km vertically. The standard deviation of ATOVS temperature data from radiosonde readings is 1.3 K.

Based on the ATOVS satellite information we compiled a data archive that contained temperature values at isobaric levels of 1000-100 hPa in a disturbed atmosphere over the Sulawes seismically active region.

Based on the methodology for analyzing vertical temperature profiles described in [13, 14] in this work we studied variations in temperature, pressure and humidity in a perturbed atmosphere starting from the surface layer (~ 0.5 km) to the lower stratosphere (~ 16 km). Separate temperature profiles at isobaric levels of 950 hPa (~ 0.5 km), 350 hPa (~ 8.0 km), 250 hPa (~ 10.0 km) and 100 hPa (~ 16.0 km) are presented in Fig. 2. The arrow indicates the earthquake date M = 7,5. On the horizontal axis, are plotted the measurement moments in increments of 3 hours from September 15 to October 7, 2018.

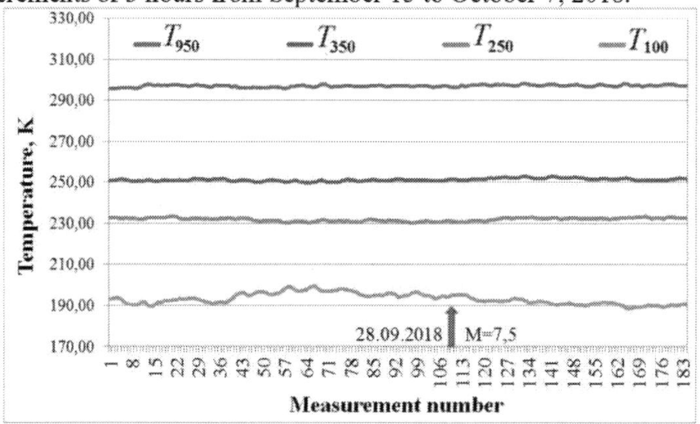

Fig. 2. Variations of temperature profiles at isobaric levels of 950, 350, 250 and 100 hPa over the focal region of the Sulawes earthquake.

As can be seen from Fig. 2 graphs of the temperature series T during the preparation of earthquakes during the main shock and during the aftershock activity in the troposphere and lower stratosphere do not change significantly and reflect the uniform temperature regime of the equatorial zone of Indonesia.

Using correlation analysis estimates of consistent temperature variations on certain isobaric surfaces above the focal region of a strong earthquake were obtained and correlation coefficients R were calculated in the troposphere above the center of the Sulaweski earthquake between the temperature series: T_{950} and T_{350} at isobaric levels of 950 and 350 hPa, T_{350} and T_{250} per levels 950 and 250 hPa, T_{950} and T_{250} at levels 950 and 250 hPa, the coefficient R is positive and varies from 0.36 to 0.61.

Consequently, in a disturbed atmospheric layer from 0.5 to 10 km, temperature variations as a reaction to the influence of geophysical factors occur in a coordinated manner. The values of the coefficient R between the temperature series T_{950} and T_{100} at isobaric levels of 950 and 100 hPa, T_{250} and T_{100} at levels of 250 and 100 hPa are negative equal to -0.25 and -0.8, respectively. The results obtained are consistent with the conclusions of [14].

To study the temperature field anomalies above the seismically active region focusing point during the geodynamic activity the method of constructing difference maps described in [13] was used. The methodology included the selection of data, the calculation of the temperature difference T of the perturbed and unperturbed atmosphere and the construction of a map of the anomalous temperature field.

For the undisturbed atmosphere temperature data were selected when the seismic background in the studied region is calm. The conditions of the disturbed atmosphere are characterized by the temperature field T during the intensification of seismic processes.

Examples of the spatio-temporal representation of temperature field anomalies at the isobaric level of 950 hPa for September 20 (*1*), September 28 (*2*) and October 1 (*3*) are shown in Fig. 3, *a*. In Fig. 3, *a* and Fig. 3, *b* the dot indicates the epicenter, the solid ellipse is the focal region, the dotted ellipse is the preparation zone. Spatial-temporal images of temperature anomalies at the isobaric level of 100 hPa for September 20 (*1*), September 28 (*2*) and October 1 (*3*) are shown in Fig. 3, *b*.

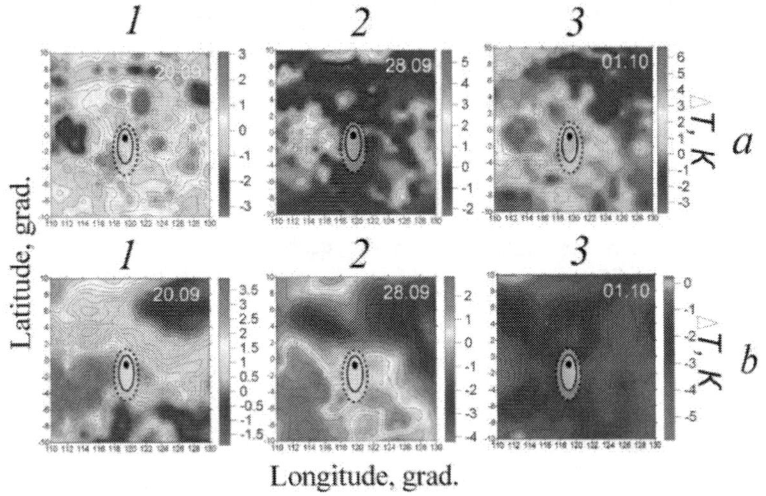

Fig. 3. Difference maps of temperature field anomalies at the levels of 950 (*a*) and 100 hPa (*b*) September 20 and 28, and October 1.

As can be seen in Fig. 3, *a* and at the level of 950 hPa above the seismically active region of the earthquake M = 7,5 (including the epicenter, focal region and preparation zone) on September 28 temperatures are below average. The focus at this level is located between two abnormal areas with elevated temperatures. Inside the focus the temperature T is below average. An increase the parameter T in the earthquake preparation zone occurred in early October.

At the isobaric level of 100 hPa in the lower stratosphere (Fig. 3, *b*) over the studied seismically active region the projection of the focus is also between two anomalous regions with elevated temperatures. An increase in T in the earthquake preparation zone at this level occurred in early October.

To detect temperature anomalies from remote sensing data we used the method for determining the temperature anomaly index, considered in [15].

In our work disturbances were detected in the temperature field over the seismically active region under study and a modified anomaly index δT was calculated for temperature series at the most informative isobaric levels, taking into account the moment of the strongest earthquake and the studied time interval [16]. These levels were selected by the maximum correlation coefficient R_{max}. Between the temperature series T_{950} at the isobaric level of 950 hPa and T_{350} at the level of 350 hPa R'_{max} = 0.56. Between the temperature series T_{250} at the isobaric level of 250 hPa and T_{100} at the level of 100 hPa R''_{max} = -0.8. The temperature profiles over the epicenter region of the earthquake with M = 7,5 for 9 hours, the time closest to the moment of the main tremor, and for the period from September 21 to October 4 in 2017 were chosen as background values.

The graph of changes in the calculated index δT for the time interval of the active existence of the Sulawes phenomenon on isobaric surfaces of 950 and 350 hPa is shown in Fig. 4. The arrow indicates the date of the Indonesian earthquake.

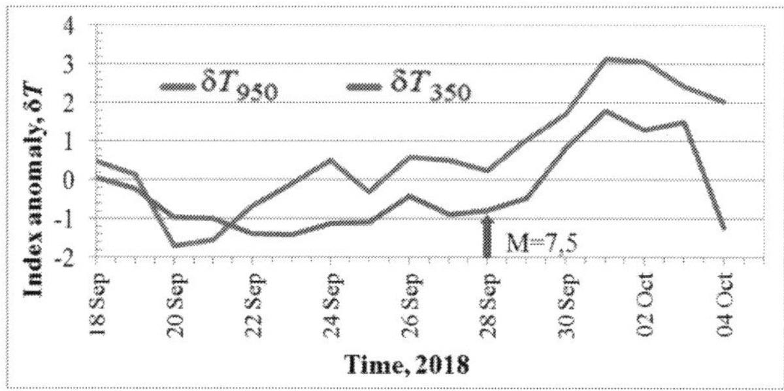

Fig. 4. Variations of the δT index on isobaric surfaces of 950 and 350 hPa.

As can be seen in Fig. 4 the decrease in the δT at 950 hPa (δT_{950}) in the surface layer of the atmosphere at an altitude of ~ 0.5 km occurred on September 20, 8 days before the main shock of the strongest Indonesian earthquake. Index δT at the level of 350 hPa (δT_{350}) in the atmospheric layer at an altitude of ~ 8 km decreased after 2 days, September 22. A consistent increase in the indices δT_{950} and δT_{350} is observed after the earthquake with M = 7,5. The parameters δT_{950} and δT_{350} reach their maximum values only by October 1. The correlation coefficient between the series δT_{950} and δT_{350} is 0.78.

In Fig. 5 shows a graph of the change in the calculated anomaly index δT for the studied time interval at isobaric levels of 250 and 100 hPa. The arrow shows the date of the earthquake with M = 7,5.

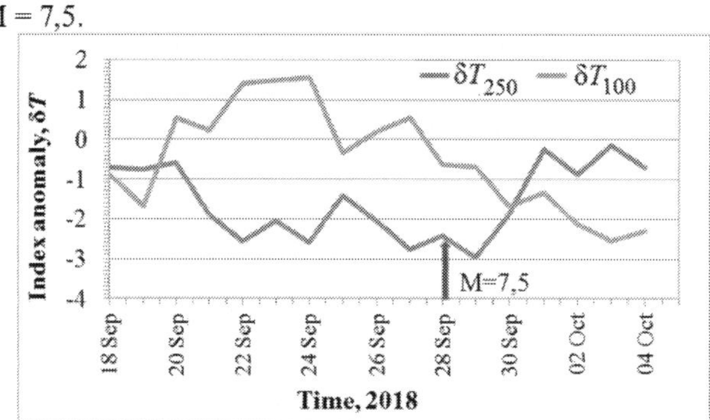

Fig. 5. Variations of the δT index on isobaric surfaces of 250 and 100 hPa.

From Fig. 5 shows that the index δT at the level of 250 hPa (δT_{250}) in the atmosphere layer at an altitude of ~ 10 km decreased after September 20, 8 days before the main moment of the strongest earthquake. The index δT at the level of 100 hPa (δT_{100}) in the atmospheric layer at an altitude of ~ 16 km also increased after September 20. A decrease in the index δT_{100} is observed on September 27, a day before the earthquake with M = 7,5. The increase in the index δT_{250} occurs by October 1. The correlation coefficient between the series δT_{250} and δT_{100} is -0.67.

4 Conclusion

The results showed that the application of the developed methodology for analyzing atmospheric disturbances during seismic activity allows us to identify temperature anomalies in the equatorial zone. It was found that in the troposphere above the epicentral region of the Indonesian earthquake with M = 7,5, anomalies with lower temperature values are observed and thermal anomalies with high temperature values are formed in the lower stratosphere during the study period.

References

1. J.R. Patton, S. Toda, R. Stein, V. Sevilgen, URL: http: //temblor.net/earthquake-insights/the-palu-koro-fault-ruptures-in-a-m7-5-quake-in-sulawesi-in.
2. H. Bao, J.-P. Ampuero, L. Meng, E. J. Fielding, C. Liang, C.W.D. Milliner, T. Feng, and H. Huang, Nature Geoscience, **12** (3) (2019).
3. C. Vigny, H. Perfettini, A. Walpersdorf, A. Lemoine, W. Simons, et al. J. Geophys. Res. **107** (B10. 2231) (2002).
4. A. Socquet, W. Simons, C. Vigny, R. McCaffrey, C. Subarya, D. Sarsito, B. Ambrosius, and W. Spakman, J. Geophys. Res. **111** (B08409) (2006).
5. USGS. URL: https://earthquake.usgs.gov
6. A. Socquet, J. Hollingsworth, E. Pathier, M. Bouchon, Nature Geoscience, **12** (2019).
7. IRIS. URL: https://ds.iris.edu/ds/nodes/dmc/data/types/events/catalogs/
8. H. Kanamori, E.E. Brodsky, Rep. Prog. Phys. **67** (2004).
9. Global Volcanism Program. Smithsonian Institution. URL: http://volcano.si.edu/volcano.cfm?vn=266030
10. C. G. Newhall, S. Self, J. Geophys. Res. **87** (C2) (1982).
11. NOAA. [Электронный ресурс]. – Access mode: URL: http://www.arl.noaa.gov/ready/cmet.html.
12. G. Goodrum, K. Kidwell, W. Winston, *NOAA KLM user's guide.* National Ocean and Atmosphere Administration. URL: http://www2. ncdc.noaa.gov/docs/klm/.
13. V. B. Kashkin, T.V. Rubleva, R.V. Odintsov, Journal of Siberian Federal University. Engineering and Technologies, **11** (8) (2018).
14. R. V. Odintsov, V. B. Kashkin, T. V. Rubleva, A. A. Romanov, I. V. Alekseenko, Sat. materials of the V International Scientific Conference "Regional Problems of Remote Sensing of the Earth" (September 11-14, 2018, Krasnoyarsk), Krasnoyarsk, Siberian Federal University (2018).
15. Yu.I. Shokin, N.N. Dobretsov, E.A. Mamash, V.A. Kikhtenko, P.V. Voronina, V.V. Smirnov, D.L. Chubarov, Computational technologies, **20** (5) (2015).
16. V. B. Kashkin, R. V. Odintsov, T. V. Rubleva, A. A. Romanov, K. V. Simonov, Trigger effects in Geosystems (Moscow, June 4-7, 2019): Materials of the V International Conference / Ed. V.V. Adushkina, G.G. Kocharyan, IDG RAS, M. TORUS PRESS (2019).

Anthropogenic impact on the air environment condition of megalopolis and suburban areas of forestland

Vadim Rogov[1], *Nataliya* Kravchenko[2], *Oksana* Bagaeva[1], *Anastasiya* Guseva[3],and *Evgenii* Sugak[1]

[1]Reshetnev SibSU, Life Safety Department, Krasnoyarsk, Russia
[2]Reshetnev SibSU, Construction Materials Engineering and Wood Science Department, Krasnoyarsk, Russia
[3]ReshetnevSibSU, Foreign Language Department, Krasnoyarsk, Russia

Abstract. The anthropogenic impact of the megalopolis on the release of volatile matters by woody plants, air ionization in the suburban areas of the forestland for the period 1998 - 2018 was monitored. A decrease in the emission of volatile matters by woody plants, negative ions, an increase in positive ions was revealed. The composition of the air environment of a megalopolis for a long time was studied. It has been established that anthropogenic factors (dust content, gas contamination, electromagnetic radiation) increase the amount of positive ions by order of magnitude. It was revealed that the presence of electromagnetic radiation significantly affects the physical properties of air. The effect of ionization and volatile matters on the deposition rate of suspended particles was determined in laboratory conditions. Theoretical studies have been carried out, on the basis of which deterministic, differential equations of the effect of ionization on dust content, gas contamination of the air environment and the interaction of air ionization with polydisperse dust, volatile substances to produce adsorbate and its deposition rate have been obtained. The data of experimental and theoretical studies made it possible to develop measures to reduce anthropogenic environmental factors, which will improve the ecological situation and reduce the incidence of megalopolis residents.

1 Introduction

Suburban and park areas are the lungs of a metropolis, affecting the cleaning of its air environment. A positive effect is achieved due to the release of volatile matters and negative ions. One of the reasons for the increase in the incidence of the inhabitants of the metropolis is its anthropogenic impact, which has a negative effect on the physical and chemical properties of the air environment of the suburban areas of the forestland.

Staying in the forest mobilizes the body's defenses and enhances its immunobiological functions. Volatile plant emissions contribute to the accumulation of vitamin C and other biologically active substances in it [1]. There is a significant amount of negative ions in the forest air. Inhalation of ionized air leads to decrease in sugar and increase in oxygen in the blood, increases the activity of enzymes.

Until the present, studies and monitoring of the influence of anthropogenic factors on the emission of volatile matters and natural air ionization by woody plants have not been conducted. The monitoring results showed decrease in volatile matters by woody plants and increase in the amount of positive air ions due to urbanization and the growth of anthropogenic factors.

The aim of the work is to study the influence of anthropogenic impact on the release of volatile matters and natural air ionization in the suburban areas of the forestland, their interaction. The paper studies the physical and chemical properties of forest air - the presence of negative and positive ions, the release of volatile matters by woody plants. The studies were conducted during the period 1998-2018. These studies allow us to determine the influence of the anthropogenic impact of the metropolis on the tropobiosphere of coniferous forest.

Theoretical dependences of the effect of natural ionization on the concentration of dust and the pathogenic microflora of the air environment were obtained.

A comparative analysis of published materials on the emission of volatile matters by woody plants was carried out, as well as measurements of the natural ionization of air in the forestland and megalopolis: the amount of negative and positive ions. [1, 2, 3, 4].

2 Monitoring of the environment

The dynamics of changes in the quantitative content of negative and positive ions in the suburban areas of the forestland during the period 1998-2018 is presented in Table 1. Air ionization was measured using the ion counter device in the warm period of the year (in July), since this period is characterized by the largest amount of volatile matters emissions and an increased content of negative ions [6].

Table 1. Dynamics of changes in the quantitative content of negative and positive ions during the period 1998–2018 in the suburban areas of the forestland

Period	1998	2003	2008	2013	2018
The amount of negative ions, e/cm^3	10 000	9 000	9 000	9 200	7 000
The amount of positive ions, e/cm^3	50	150	200	300	350

From table 1 it is seen that over the indicated period in the suburban areas of the forestland, the number of negative ions decreased by 30%, the amount of positive ions increased by 7 times.

The objectives of the study also included studying the composition of the air environment in a metropolis. Data on the amount of negative and positive ions in the air environment of a megalopolis (Central and Kirov districts of Krasnoyarsk) are presented in Table 2.

Analysis of the data in table 2 shows that the amount of positive ions exceeds the amount of negative ions in the studied areas of the city. During this period, there is a significant increase in the amount of positive ions and a decrease in negative ones.

Table 2. Dynamics of changes in the quantitative content of negative and positive ions during the period 2016-2019 in Krasnoyarsk (warm season)

Наименование/period	2016	2017	2018	2019
The amount of negative ions, e/cm^3 (Central district)	500-600	500-600	300-350	300
The amount of positive ions, e/cm^3 (Central district)	700-800	1100-1200	1900	2200
The amount of negative ions, e/cm^3 (Kirov district)	1100	1100	900	850-900
The amount of positive ions, e/cm^3 (Kirov district)	900	1200	1800-1900	2100

According to the observations of the authors, electromagnetic radiation (EMR), which is found in almost all districts of the city, has a negative effect on the inhabitants of the city and the air environment. The presence of electromagnetic radiation negatively affects the physical properties of the air environment, increasing the amount of positive ions by an order of magnitude compared to negative ones, forming clusters of dust and gases [5].

The authors' studies established that the air in forestland and meadows contains from 2,000 to 10,000 negative aero ions in cm^3 (e/cm^3), and in residential buildings their concentration decreases to 25 e/cm^3, and in residential districts of the city it does not exceed 500 e/cm^3 or is completely absent and an excess of positive ions is observed, which adversely affects the health of the population of the metropolis.

Based on the studies and the state report on the condition and environmental protection in the Krasnoyarsk Territory during 2016-2018 the data were obtained which are presented in table 3.

Table 3. The dynamics of the content of harmful substances in the air of a megalopolis during 2016-2018

	Suspen-ded matters	Sulfur dioxide	Carbon oxide	Nitrogen dioxid	Nitrogen oxide	Hydrogen sulfide	Benza-pyrene
2016	0,75(4,8)	(3,0)	0,4 (1,8)	0,85 (1,15)	0,4(-)	(15,1)	5,1(40,6)
2017	0,85(6,0)	0,26 (3,4)	0,45 (3,0)	0,75 (1,21)	0,35 (1,01)	(14,1)	5,3(40,1)
2018	0,89(7,6)	0,28 (4,02)	0,47 (3,6)	0,9 (1,5)	0,25 (1,79)	(14,8)	6,5 (50,6)

During the studied period there is an increase in suspended particles and harmful substances. In authors' opinion, this is due to an increase in the number of automobile transport in the metropolis.

3 Experimental research

To determine the effect of negative ions on the content of dust and gas in the air environment, original studies were conducted in laboratory conditions. A laboratory setup was developed and mounted, which includes a source with control of the amount of negative ions and biologically active substance (BAS). BAS was a light fraction of essential oils of woody plants (spruce, pine, cedar, juniper, forbs). The results are presented in Figure 1.

The figure shows the time dependences (t, m) of the concentration of suspended dust particles (*C, mg/m^3*) in air without additives (curve 1), in an air-ionized environment (curve 3), and with an admixture of volatile productions (curves 2 and 4).

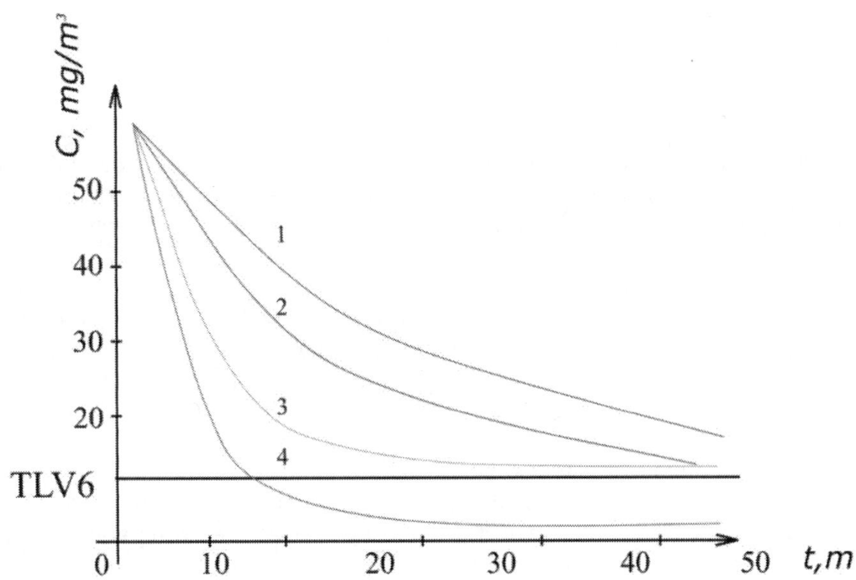

Fig. 1. The dependence of dust concentration on the type of air treatment:

1 - natural dust deposition;
2 - under the influence of volatile substances;
3 - under the influence of ionization;
4 - under the influence of ionization and volatile substances.

Studies have shown that negative ions in combination with volatile substances have an effective influence on reducing of dust content and gas contamination of the air environment.

4 Theoretical description of the process

To describe the impact of anthropogenic factors on air ionization, a theoretical study was carried out and mathematical dependences were obtained.

The model equilibrium state of a layer of adsorbed ions on the particle surface is described by the Fokker-Planck equation[6]:

$$\frac{\partial p}{\partial t} = \frac{\partial}{\partial x_i}\left(i_i p - \frac{\partial}{\partial x_i} D_{ij} p \right) \tag{1}$$

where: $x_1 = x$, $x_2 = y$, $x_3 = z$ are the coordinates of the point in the volume field of the particle-adsorbate conglomerate containing the particle-medium interface, l_i are the projections of the mobility vector of adsorbed ions corresponding to them, p is the charge distribution density; D_{ij} is the tensor of their diffusion coefficients.

After converting the Fokker-Planck equation (1), taking into account the properties of the adsorbate material (dust particles), we obtain the equation[6]:

$$\frac{\partial p}{\partial t}=D\frac{\partial^2 p}{\partial r^2}+\alpha\frac{\partial U}{\partial r}\frac{\partial p}{\partial r} \tag{2}$$

where r is the reduced radius of the adsorbate, α is the mobility vector.

The interaction of air ions with the volatile substances of woody plants affects the structure of the adsorbate with the formation of polydispersed moments with the further aggregation of aerosol solid particles and their deposition from the air environment.

If it contains only water and ions, the adsorbent-adsorbate interface will coincide with the active surface of the aerosol particles. If the medium contains a high molecular weight additive, for example, volatile matters, a layer such as an adhesive coating will eventually be formed on the surface of the particle. In the first case, the total mass of particles is limited in time, the mass of the adsorbate will increase to an equilibrium state. Then, due to an increase in molecular exchange with the medium, the outer interface of the adsorbate is eroded, and its mass ceases to increase over time. In the second case, the thickness of the adhesive film is not limited in time. The particle becomes two-layer. Moreover, such a conglomerate structure will determine the nature of the adsorption of water and ions. The adhesive film may be hydrophobic, which completely excludes the adsorption of water. In the case of its hydrophilicity, the formation of a third layer of water and ions is also possible.

In a practical sense, the mechanism of deposition of aerosol particles is represented by translational degrees of freedom. In this case, we mean polydisperse particles suspended in air, because otherwise, the probability of collision of particles moving in the same direction is low. The aerosol mixture on the average particle size scale is generally heterogeneous, that is, there are density fluctuations and randomly directed microflows in the medium (convection, thermal diffusion, diffusion). Particles of high dispersion are carried away by microflows and are pushed by density fluctuations.

All of the above leads to the random movement of particles in the Brownian picture. Moreover, as it is easy to show, the Stokes law is its determining factor [6]:

$$\mathrm{Fc}=6\pi\eta\,\mathrm{r}\mathrm{v}_{ч}, \tag{3}$$

where Fc is the force of viscous friction, η is the dynamic viscosity of the medium, r is the radius of the particle, $v_ч$ is its velocity.

To precipitate the obtained adsorbate in air environment under the influence of ionization and volatile matters, the particle motion in the field of gravity was considered. At the initial moment t=0 a spherical particle with radius r moves under the action of gravity in a viscous medium with a velocity v_0. Two opposing forces act on it: the gravity mg and Stokes $6\pi\eta\,\mathrm{r}\mathrm{v}_ч$. The resulting or acting force mg- $6\pi\eta\,\mathrm{r}\mathrm{v}_ч$ according to the d'Alembert principle is equal to the product of mass and acceleration:

$$mv_ч = mg - 6\pi\eta r v_ч \tag{4}$$

After transformations of equation (4), we obtain the final expression for determining the deposition rate of adsorbate particles V_{0c}, which is determined by the formula

$$v_{0c}=\frac{\mathrm{mg}}{6\pi\eta\, r}. \tag{5}$$

The obtained mathematical dependences explain the mechanism of interaction of air ionization with polydispersed dust, volatile matters to produce adsorbate and the rate of its deposition from the air environment

5 Conclusions

1. For the period from 1998 to 2018, studies were conducted on the release of volatile matters by woody plants under the forest canopy in the suburban areas of the forestland (50 - 70 km from the metropolis). As a result it turned out that the amount of negative ions decreased over the indicated time period by 30%, while the amount of positive ions increased sevenfold.

2. For the period from 2016 to 2019, studies of the air environment of the metropolis were conducted. An increasing dynamics of positive ions and a decrease in negative ions were revealed. The amount of suspended particles and harmful substances in the air increased due to an increase in road transport.

3. The effect of electromagnetic radiation on the increase in positive ions by an order of magnitude compared with negative ones has been established.

4. Studies have been carried out at the developed laboratory facility, which showed that negative ions in combination with volatile matters have an effective impact on reducing dust content and gas contamination of the air environment.

5. The obtained mathematical dependences provide a theoretical justification for the interaction of negative ions and volatile matters with polydisperse dust in the air.

Based on the theoretical and experimental studies, measures have been developed to reduce the anthropogenic impact on the air environment and residents of the megalopolis:

1. To create air with the content of negative ions up to 8000-10000 e/cm^3 in the park zones of Krasnoyarsk.

2. Creation of an ecologically clean zone in parks and squares of Krasnoyarsk by planting phytoncide woody plants.

References

1. V.A. Rogov, R.A. Stepen, V.V. Kulak, Conifers of the Boreal Area: Theor. and Appl. Research J., **33,** 168-170 (2015)
2. V.A. Rogov, A.V. Rogov, R.A. Stepen, Chemistry of plant raw materials, **4,** 181 – 184. (2008)
3. Luis A. Gurovich, Electrophysiology of Woody Plants. Electrophysiology – From Plants to Heart, Dr. Saeed Oraii (Ed.), InTech (2012)
4. Perveen, Shagufta, Chemical composition and antimicrobial activities of essential oils of some coniferous plants cultivated in Egyp. IJPR, **16** (Jan. 2017)
5. V.A. Prusakova, V.A. Rogov, V.V. Kulak, Forest and chemical complexes; problems and solutions, **1,** 206 – 208 (2015)
6. V.A. Rogov, *Influence of negative ions and volatile terpenoids on the air purification of processing areas of woodworking enterprises* (Publishing House of Moscow State University, 2002)

AUTHOR INDEX